# NEW

# ASTRONOMER

# NEW ASTRONOMER

## Carole Stott

*Editorial Consultant*
**Amie Gallagher**
*American Museum of Natural History,*
*Hayden Planetarium*

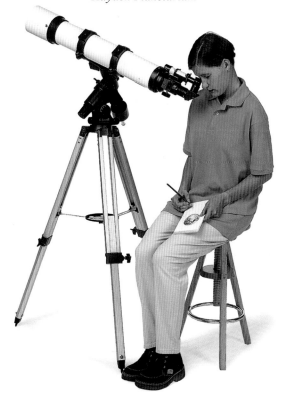

## DK PUBLISHING, INC.

NEW YORK

www.dk.com

### A DK PUBLISHING BOOK

**Senior Editor** Heather Jones

**Project Art Editor** Vanessa Hamilton

**US Editor** Alrica Green

**Designer** Joanne Long

**DTP Designer** Jason Little

**Picture Research** Andrea Sadler

**Production Controllers** Michelle Thomas, Silvia La Greca

**Senior Managing Editor** Martyn Page

**Senior Managing Art Editor** Bryn Walls


First American Edition, 1999

2 4 6 8 10 9 7 5 3 1

Published in the United States by DK Publishing, Inc.,
95 Madison Avenue, New York 10016
www.dk.com
Copyright © 1999 Dorling Kindersley Limited
Text copyright © 1999 Carole Stott


For Mum, Dad, Jane, Martin, and David

Library of Congress Cataloging-in-Publication Data

Stott, Carole.
    The new astronomer / by Carole Stott.--1st American ed.
        p.   cm.
    Includes index.
    ISBN 0–7894–4175–6 (alk. paper)
    1. Astronomy -- Observers' manuals.  2. Astronomy
    --Amateurs' manuals      1. Title.
QB64.S756 1999
520--dc21                                  98–45283
                                                    CIP
Reproduction by GRB, Verona, Italy

Printed in China by Toppan Printing Co.,
(Shenzhen) Ltd.

# CONTENTS

## INTRODUCING THE NIGHT SKY

## ASTRONOMICAL TOOLS AND TECHNIQUES

**Sweeping the sky with binoculars**

Solar eclipse

Galaxy M33 in Triangulum

The planet Saturn

The Pleiades star cluster

Globular cluster M10 in Ophiuchus

Aurora Borealis

WHEREVER WE ARE ON EARTH WE CAN LOOK UP AND SEE SKY. THE SKY IS CONSTANTLY CHANGING, AND OUR PARTICULAR VIEW OF IT DEPENDS ON LOCATION. THIS SECTION CONSIDERS WHAT WE CAN SEE FROM WHERE, HOW OUR VIEW ALTERS, AND HOW SKY MAPS CAN HELP US FIND OUR WAY AROUND.

# INTRODUCING *the* NIGHT SKY

# LOOKING UP

When we look up at the sky, we are looking out into the vastness of space. In every direction there are countless billions of stars. There are entire galaxies full of stars, and clouds of dust and gas left behind by exploding stars. Nearer home is our own star, the Sun, that gives us light and heat on Earth.

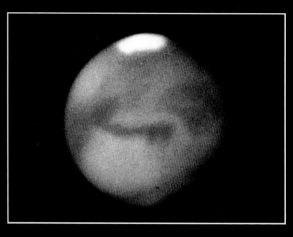

### THE MILKY WAY

On very dark nights, a band of light can be seen across the sky. This is our view into our galaxy. There are so many stars that it looks like spilled milk, thus its name, Milky Way.

### THE PLANETS

The planets look like bright stars in the sky but, on closer inspection, they take on a disk shape. Mars (above) is a spinning ball of rock with polar caps and dark patches, all of which can be seen from Earth with a telescope.

### DISTANT GALAXIES

Galaxies are vast collections of stars; they are so distant that even one of the closest, the Andromeda Galaxy, appears in the sky as a small blur of light. With a telescope, the shape of the galaxy can be seen.

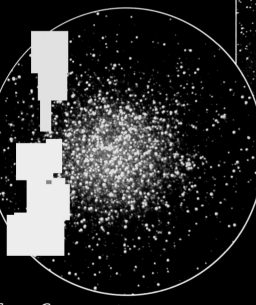

### STAR CLUSTERS

Some stars live in clusters with thousands of other stars. This cluster consists of old stars that have been together since they were born.

## STAR PATTERNS

When we look at the night sky, our eyes are drawn to the most brilliant stars. These dots of light make familiar patterns, like that of Orion, the Hunter. They can be recognized over and over again and used by observers to find their way around the sky.

## THE DEATH OF A STAR

Every star is unique because each is in a different stage of life. The star above has started to die and is throwing off huge shells of gas.

## BRIGHT NEBULAE

Nebulae are clouds of gas and dust that are formed from material thrown off by the stars. They produce new stars that will shine in Earth's sky in millions of years' time.

## ECLIPSE OF THE SUN

From time to time a rare event can be seen in the daytime sky. This is an eclipse of the Sun and it occurs as the Sun's light is blocked by the Moon.

9

# OBSERVATORY EARTH

There are tens of thousands of celestial objects existing in the sky above us. When these objects are viewed from Earth, they appear as if on a flat screen an equal distance away from us. In fact, all celestial objects are at different distances away from Earth and also at vastly different distances away from each other. The view from Earth is also misleading in terms of the size and nature of celestial objects. A remote galaxy consisting of billions of stars appears the same size as a single star in the sky. Planets are easily mistaken for stars, too, and a fuzzy patch can be one of several things: a cluster of thousands of stars, a nebula, or a passing comet. In time, it is possible to become familiar with many different objects.

### PLANET EARTH

When we look out into space from Earth, all the objects we see in the sky appear to be moving around us. In reality it is Earth that is turning and moving through space.

### The Moon
*One of the smallest celestial objects, the Moon appears largest in the sky because it is the closest object to Earth.*

### The planet Jupiter
*More distant than the Moon are the planets. Jupiter looks like a small bright star. About 70,000 Moons could fit across the diameter of Jupiter. It looks small because it is about 2000 times further away.*

### YOUR WINDOW ON THE UNIVERSE

As long as the sky is clear, it is possible to look out into space. Whatever country an observer views the sky from, in no matter which direction—north, south, east, or west—a window is opened onto the universe. Through this window can be seen objects of vastly different sizes and structures, at widely varying distances from Earth.

*A nebula is formed from material thrown off by dying stars and, over millions of years, produces new stars.*

*All stars start life as part of a cluster. Some, like these, stay together; others drift apart.*

*A comet's tail is typically six million miles long—a bright comet such as this one can be seen about once every ten years.*

**Galaxies**
*The largest and furthest objects to view are the far-off galaxies, lying beyond our own. They look tiny because they are so distant, but they can contain billions of stars.*

**The stars**
*Each star is a huge globe of hot, luminous gas, like the Sun.*

**The Veil Nebula**
*In the space that exists between stars are giant clouds of gas and dust. These are called nebulae.*

**Comet West**
*Comets are dirty snowballs at the edges of the Solar System. When a comet comes close to Earth, it appears as a spectacular object in our sky*

*Only one of these objects is a star—the others are galaxies, never seen until 1995.*

## MEASURING DISTANCE

Distances within the Solar System are measured by radar. To measure the distance to nearby stars, the parallax method (right) is used. A star is observed at opposite sides of Earth's orbit, six months apart. The apparent shift in the star's position is used to calculate its distance. Remote stars are measured by the light they emit.

*Apparent shift —the parallax— of the star.*

*Position of Earth in July*

Sun

Star A

Star B

*Position of Earth in January*

*The more distant the star, the smaller the parallax.*

## BEYOND OUR PRESENT VISION

In the universe there are countless galaxies that we have never seen. In 1995, the powerful Hubble Space Telescope focused on a small patch of sky about a thirtieth of a Full Moon in diameter for a period of ten days and produced the picture above. Astronomers have been able to estimate that there are more than 3,000 galaxies visible within this image.

# THE CHANGING SKY

Our home planet, Earth, travels through space; it spins around once a day and takes one year to complete an orbit of the Sun. Although we cannot feel Earth's motion, we can perceive it by observing the movement of celestial objects across the sky. The Sun's daily movement is the most familiar to us, but the stars also move in a nightly progression across the sky. Earth's yearly movement around the Sun is detected by changes in the Sun's position in the sky: it is high in summer and low in winter. Similarly, as Earth moves through space against the backdrop of stars, different stars take their places above our heads during each succeeding month.

*The Sun climbs high in the morning sky.*

**Earth turns** anticlockwise as viewed from above the north pole.

**Earth's daily spin**
*Earth spins on its own axis once every 24 hours. As it rotates, we experience alternate periods of day and night, or light and darkness, on Earth, as we first face towards, then away from the Sun.*

23.5°

*The axis remains inclined at the same angle to the vertical.*

## DAILY CHANGES

As Earth makes its daily spin, first turning toward the Sun, the Sun appears to rise above the eastern horizon and to move slowly across the sky, reaching its highest at a midpoint on its path. As Earth gradually turns away from the Sun, the Sun appears to set below the line of the western horizon.

## YEARLY CHANGES IN THE NIGHT SKY

Most observers around the world see different stars in their sky during the course of a year. Observers located at the north and south poles are the exception. The order of stars seen in the course of one year is followed for each subsequent year. An observer in the northern hemisphere at a particular location at a fixed time of night, looking in the same direction, can see how the sky changes.

**Mid-January 10.30 p.m.**
*Orion (p.112) is close to the horizon (right). As the night progresses, Canis Major (p.113), with brilliant Sirius, will move across the sky.*

**Late May 10.30 p.m.**
*Bright star Arcturus (lower right) shines out in Boötes (p.120). The semicircle of bright stars (center left) is Corona Borealis (p.120).*

**Mid-September 10.30 p.m.**
*Toward year's end, Aquarius (p.108) is in the sky, and the other zodiacal constellations will then appear in turn throughout the year.*

*The Sun is highest in the middle of the daylight hours.*

*The light from the Sun brightens the sky—the distant stars are lost from view but will be seen once the Sun sets.*

**Daily changes in the night sky**
*The stars change position in the course of an evening. Although we notice their change in position, their movement is imperceptible to the human eye, but when a long-exposure photograph is taken, the recorded image clearly shows the stars' movement as trails of light.*

## YEARLY CHANGES IN THE DAYTIME SKY

It is only an approximation to say that the Sun rises in the East and sets in the West. As the yearly seasons progress, and the Earth completes its orbit around the Sun, the Sun seems to rise and set at different points along the horizon.

**Sunset, August 6**
*The Sun is about to set in a location at latitude 50°N. The photographer is positioned looking toward the western horizon.*

**Sunset, August 7**
*At the same location looking toward the western horizon at the same time a day later, the Sun is setting at a point further south.*

**Sunset, August 11**
*As the summer progresses, the Sun sets below the line of the horizon at a point that becomes further and further toward the south.*

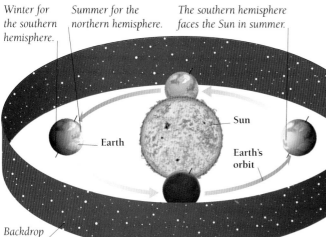

*Winter for the southern hemisphere.*

*Summer for the northern hemisphere.*

*The southern hemisphere faces the Sun in summer.*

Sun

Earth

Earth's orbit

*Backdrop of stars*

### Why the sky changes throughout the year
*Earth takes one year to complete its orbit of the Sun. As Earth follows its yearly path with the stars forming a backdrop to its movement, an observer on Earth will see progressively different stars at the same time on successive nights throughout the year.*

# YOUR LOCAL VIEW

An observer looking into space from Earth looks out onto an imaginary sphere that encompasses Earth (pp.16–17). From any position on Earth, it is only possible to see part of the sphere at any one time. Although the view changes during the course of a year, most observers will never be able to see the entire celestial sphere. The part of the sphere that is visible to you is directly related to your latitude on Earth. Observers at the same latitude around the world will see the same sky in turn. To obtain a good view of the sky above you, observers should choose the best-possible site for their observations.

## WHAT STARS ARE VISIBLE FROM WHERE

The portion of celestial sphere visible at any location on Earth is determined by latitude. As an observer moves north or south, the portion of visible sphere changes. The stars on the relevant portion of sphere are seen during the year. (These stars are always in the sky but are only visible when above the horizon at night.)

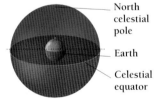

**View from the north pole**
*The north celestial pole is directly overhead. Only the stars in the northern half of the sphere are visible.*

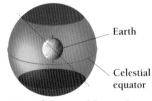

**View from mid-latitudes**
*The stars that are positioned close to your celestial pole are always in the sky. Other stars can be seen in the sky during the course of the year.*

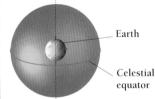

**View from the equator**
*The celestial equator is overhead and the celestial poles are on opposite horizons. All the stars are seen during a year.*

**Key to spheres**

■ Stars always visible

■ Stars visible at some point during the year

■ Stars never visible

...... Horizon

• Observer's position

## TOWN AND COUNTRY

Celestial objects such as stars and planets are seen to best advantage in a dark sky, away from street and house lights. Country observers have the most favorable views and high ground gives an even better vantage point.

Sirius

## LATITUDE AND LONGITUDE

Observers on the same line of latitude can be on opposite sides of Earth—at different longitudes—but they will see the same stars. For instance, an observer in Flagstaff, Arizona, US will see the same stars as an observer in Tokyo, Japan, but at an interval of several hours.

**Arizona, USA, 11.00 p.m.**
*The constellation of Orion is seen high above the southern horizon in the late evening sky of Arizona.*

**Tokyo, Japan 11.00 p.m.**
*Several hours later, late-evening observers in Japan can see Orion high above their horizon.*

### In the city

*The sky above a brightly lit town or city is never truly dark but it is possible to see bright stars and planets. The bright stars usually form the constellation shapes, so the city is a good place to learn the basic star patterns.*

Sirius

### In the country

*The darkest skies are in the country away from artificial lighting. A country observer can see faint stars as well as the bright ones, whereas a city observer will see only the bright stars. This means about 300 stars will be seen in the city, compared to about 3,500 from a location in the country.*

## ENHANCING YOUR VIEW

Find out when your sky is dark by using the sunrise and sunset times for latitudes around the world (p.136). Using the naked eye, you will be able to pick out hundreds of bright stars in a dark sky, although fainter objects can be more difficult to observe. In such cases, the technique of averted vision is often helpful. Do not look directly at the object but slightly away. The object will appear as an image, formed by the sensitive edge of the retina.

### Viewing methods

*The symbols below appear throughout the book. Where they are next to a picture they indicate how to obtain the view shown; when in lists they suggest how best to view an object.*

With the naked eye

With binoculars

With a telescope

With a CCD set-up

### Using binoculars

*With binoculars, 43,000 stars are bright enough to be seen. Mountains and craters on the Moon and star clusters, nebulae, and galaxies are also revealed.*

Sit on a chair to steady yourself.

## NIGHTLY STAR MOVEMENT

An observer's latitude affects what part of the celestial sphere can be seen and how stars move across the sky. The identity of the celestial pole (north or south) and its position in the sky determines how the stars move. The position of the celestial pole in your sky corresponds to your latitude figure. For example, for people at 35°N the north celestial pole is located 35° above the northern horizon.

### Star movement at mid-latitudes

*Stars appear to circle the north celestial pole in the northern hemisphere sky and the south celestial pole in the southern. A long-exposure photograph records each star as a trail of light.*

### View from mid-latitudes

*From here some stars are seen circling the celestial pole; others rise above the horizon, move across the sky, and set below the opposite horizon.*

### View from the north pole

*An observer at the north pole sees the stars circling overhead. Observers at the south pole see the stars circle in the opposite direction.*

### View from the equator

*Stars above the equator move across the sky over the observer's head. They rise above the eastern horizon and set below the western.*

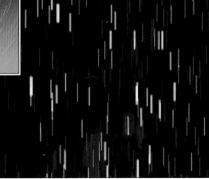

### Star movement seen at the equator

*The light from individual stars is recorded as dashes in this long-exposure photograph taken from an observing site at the equator. The dashes point down toward the horizon where the stars will eventually set.*

# MAPPING THE SKY

A sky map is an essential tool for astronomers—not just for beginners, but also for experienced observers. Just as a terrestrial map helps us to find our way on Earth, a map of the heavens helps us to navigate our way around the sky and to find particular objects. All sky maps are based on the notion of a celestial sphere, an imaginary ball that completely envelops Earth. The stars seem fixed to the inside of this sphere, staying in their relative positions as they move across the sky. Although the celestial sphere is an illusion, it is useful as a basis for making maps of the night sky.

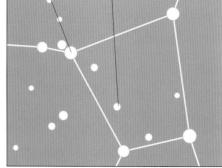

*Declination is measured north and south of the celestial equator.*

*The celestial equator is the equivalent of the equator on Earth, dividing the sphere into hemispheres.*

*Declination south of the celestial equator has a negative value. This line can be expressed as Dec −20° or 20°S.*

## THE CELESTIAL COORDINATES

The coordinates, right ascension (RA) and declination (Dec), work in the same way as latitude and longitude on Earth, to pinpoint the position of a star or planet on the sphere. RA values range from 0 hrs through to 24 hrs; Dec values range from 90° north through to 0° to 90° south.

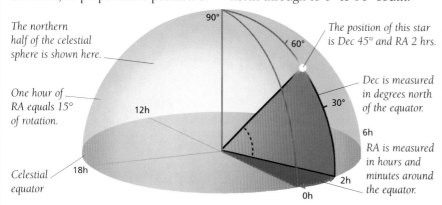

*The northern half of the celestial sphere is shown here.*

*One hour of RA equals 15° of rotation.*

*Celestial equator*

*The position of this star is Dec 45° and RA 2 hrs.*

*Dec is measured in degrees north of the equator.*

*RA is measured in hours and minutes around the equator.*

90°
60°
30°
12h
18h
6h
2h
0h

## THE CELESTIAL SPHERE

An imaginary sphere around Earth is used as a basis for all sky maps. A network of lines and a system of coordinates help observers to pinpoint specific objects on the sphere, and hence in the sky.

## STAR MAGNITUDE

The stars we see in the sky vary in brightness. Astronomers use a scale called magnitude to signify how bright a star appears to us from Earth. This is apparent magnitude and it is not the same as the star's actual brightness, or luminosity. On star maps, each star is given a magnitude number; the larger the number, the fainter the star. Stars that appear brighter than mag.1 have a zero or negative value. For example, the brightest star in the sky, Sirius, is mag.−1.46. Astronomers also use the magnitude scale for other space objects. The Full Moon is mag.−12.5, and the planet, Venus, when it is at its brightest, is mag.−4.4.

*Stars of mag.6 and brighter are visible with the naked eye.*

*This mag.3.1 star has a larger circle than the map's fainter stars.*

*This mag.5.3 star has a smaller circle than the brighter star above left.*

**Stars in the constellation Hercules**
*The stars (above) have a range of magnitudes. The faintest, mag.10, invisible to the naked eye, can be seen through binoculars.*

**Constellation map of part of Hercules**
*Circles on maps denote the stars' magnitudes; the circles are used to represent the nearest whole number within the magnitude scale.*

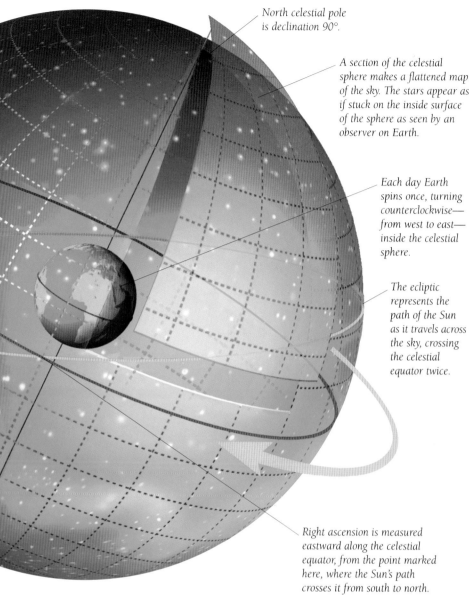

North celestial pole is declination 90°.

A section of the celestial sphere makes a flattened map of the sky. The stars appear as if stuck on the inside surface of the sphere as seen by an observer on Earth.

Each day Earth spins once, turning counterclockwise— from west to east— inside the celestial sphere.

The ecliptic represents the path of the Sun as it travels across the sky, crossing the celestial equator twice.

Right ascension is measured eastward along the celestial equator, from the point marked here, where the Sun's path crosses it from south to north.

## SKY MAPS

Sky maps are created by dividing the celestial sphere into sections and flattening them out to make several flat maps. The sphere is traditionally divided into two north and south polar and several equatorial maps.

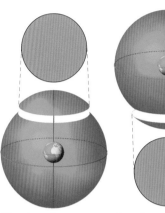

**North polar map**
*This map is 50°N to 90°N, centered on the north celestial pole.*

**South polar map**
*This map is 50°S to 90°S, centered on the south celestial pole.*

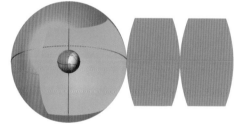

**Equatorial maps**
*A broad band either side of the equator can be divided into separate maps showing the stars that are visible throughout the year.*

## OTHER SKY MAPS

Sky maps vary in the number of objects they include and the detail they give. Some maps show faint stars or star types. Star catalogs listing celestial coordinates are available as books or disks, and are essential for use with equatorial telescopes (pp.26–7).

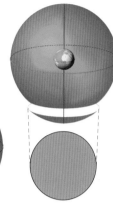

**Computer maps**
*Sky maps are available as computer software. Some have landscape locations and show the sky at certain times and dates; others show the progression of objects against the starry background.*

**Planisphere**
*A planisphere is a useful reference tool for both amateur and professional astronomers. See pp.22–3 to find out how to use the planisphere accompanying this book.*

MANY OBJECTS IN THE NIGHT SKY CAN BE SEEN WITH THE NAKED EYE, BUT INSTRUMENTS ENHANCE OUR VIEW. THIS SECTION INTRODUCES THE PLANISPHERE, DISCUSSES BINOCULARS, TELESCOPES, CCD SET-UPS, AND HOW TO PHOTOGRAPH THE SKY.

## CAMERA AND TELESCOPE

A single-lens reflex (SLR) camera can be fitted to the eyepiece end of a telescope. The camera's own lens is removed and an adaptor is used to attach the camera to the telescope. The light is gathered by the telescope lens and the camera is used simply to hold the film.

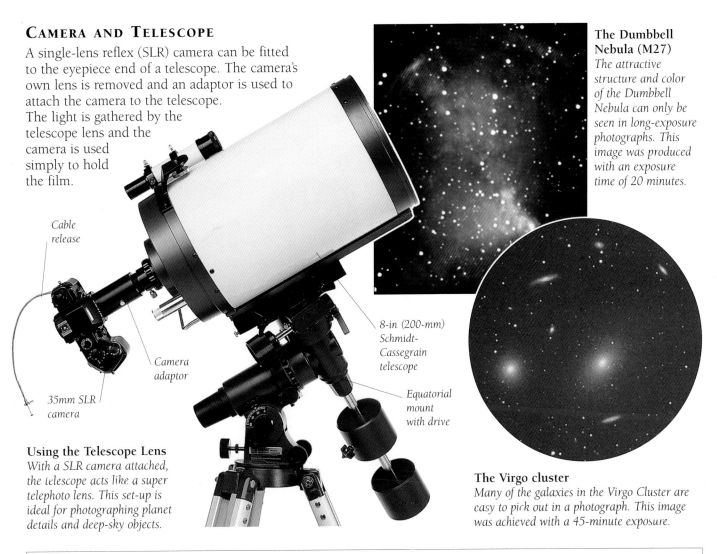

Cable release

Camera adaptor

35mm SLR camera

8-in (200-mm) Schmidt-Cassegrain telescope

Equatorial mount with drive

**Using the Telescope Lens**
*With a SLR camera attached, the telescope acts like a super telephoto lens. This set-up is ideal for photographing planet details and deep-sky objects.*

**The Dumbbell Nebula (M27)**
*The attractive structure and color of the Dumbbell Nebula can only be seen in long-exposure photographs. This image was produced with an exposure time of 20 minutes.*

**The Virgo cluster**
*Many of the galaxies in the Virgo Cluster are easy to pick out in a photograph. This image was achieved with a 45-minute exposure.*

## ACCESSORIES

There are many types of accessories that will help you produce better images. Color filters can be fitted to the telescope eyepiece or camera adaptor and are useful both for photographing planets and deep-sky objects and for just observing them.

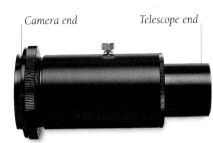

Camera end

Telescope end

**Camera adaptor**
*A specially designed adaptor that fits between the telescope and the camera can be used together with an adjustable eyepiece to allow you to achieve a range of magnifications of the image.*

Green

Blue

Red

Yellow

**Color filters**
*There are many different kinds of filters that can be fitted to the telescope eyepiece. Each filter is designed to screen a different color, or wavelength of light, and so allow other wavelengths to come through, thus highlighting details on planets, for instance.*

**Light-Pollution filter**
*Light from sodium street lighting can be screened by a light-pollution filter. The filter gives darker skies and, by contrast, brighter objects.*

**Sky with light filter**

**Sky without light filter**

**Sun filter**
*The Sun can be photographed safely with a special filter fitted over the front of the telescope.*

# DIGITAL ASTRONOMY

In recent years, digital technology has greatly expanded the amateur astronomer's horizon. Many of the spectacular images we see of celestial objects are produced by a special digital camera that uses a CCD (charge-coupled device) to convert light into digital data, to produce an image. Once used only by professionals, the CCD camera is now used by amateur astronomers. Fitted to a telescope and linked to a home computer, a CCD camera can produce vivid images of even faint objects. The home computer can also be used to log onto images recorded by the major world observatories and the Hubble Space Telescope.

## CCD IMAGES

A CCD camera is designed to obtain spectacular images of single objects, as opposed to large fields of view. It is about one hundred times more sensitive to light than the human eye, and can record objects that would otherwise be invisible. It can also resolve fine details of objects.

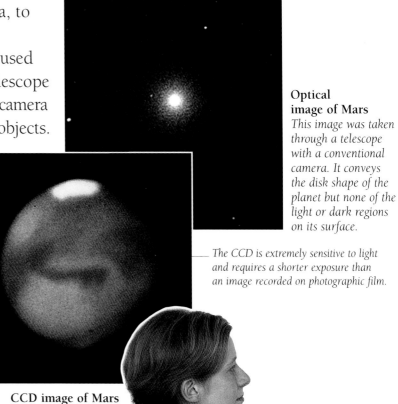

**Optical image of Mars**
*This image was taken through a telescope with a conventional camera. It conveys the disk shape of the planet but none of the light or dark regions on its surface.*

*The CCD is extremely sensitive to light and requires a shorter exposure than an image recorded on photographic film.*

## THE INTERNET

Anyone who has access to the Internet can use it to help them to explore space. An increasing number of astronomical groups now have websites. The Internet allows you to access an observatory site so that you can see what a particular telescope is looking at at any time of the night or day. Special events, such as comets and lunar or solar eclipses, can also be viewed in this way.

*Professional images can be accessed from the Internet.*

**CCD image of Mars**
*This image was taken with a CCD camera attached to a telescope. It has recorded the light and dark surface markings on Mars as well as a polar region.*

*The image can be adjusted with normal computer controls.*

## CCD SET-UP

A CCD set-up has three parts: a telescope with equatorial mount and drive (pp.26–7); a CCD camera linked to a driver module; and a computer. Software for recording and storing data is usually included with the CCD equipment.

## IMAGE ENHANCEMENT

CCD images can be viewed and enhanced by using a computer-imaging program. The image on the screen can be manipulated, for example, by altering the brightness, increasing the contrast, and adjusting the color. Imaging programs vary; some allow you to increase a particular detail but only at the expense of losing detail in other parts of the image. Others allow you to manipulate the image so that every part of the picture is enhanced.

*Three separate images are produced using the CCD set-up.*

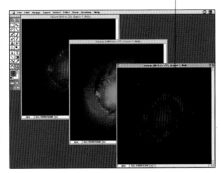

*Color scale is used to adjust colors.*

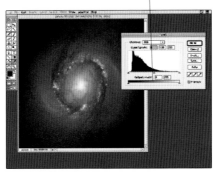

*Fully adjusted image of galaxy M83 in Hydra.*

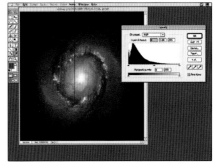

**1 Original exposures**
*Color CCD images can be built up one color at a time. These were made one at a time using three different color filters.*

**2 Combining the color images**
*Each of the three different images are manipulated before being combined. Then the combined image is adjusted for clarity.*

**3 Filtering and enhancing the image**
*The final full-color image of M83 is displayed on the screen. It has detail and depth.*

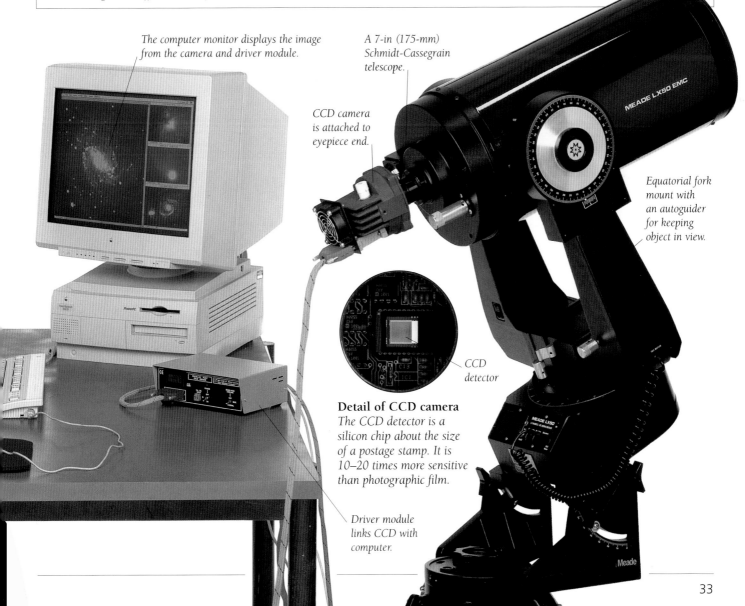

*The computer monitor displays the image from the camera and driver module.*

*A 7-in (175-mm) Schmidt-Cassegrain telescope.*

*CCD camera is attached to eyepiece end.*

*Equatorial fork mount with an autoguider for keeping object in view.*

*CCD detector*

**Detail of CCD camera**
*The CCD detector is a silicon chip about the size of a postage stamp. It is 10–20 times more sensitive than photographic film.*

*Driver module links CCD with computer.*

EARTH'S CLOSEST NEIGHBORS IN SPACE ARE THE OBJECTS OF THE SOLAR SYSTEM. THE SUN ILLUMINATES THE MOON AND PLANETS THAT SHINE AT NIGHT. THIS SECTION DESCRIBES THE PLANETS, MOON, SUN, COMETS, AND ASTEROIDS AMONG OTHER OBJECTS, AND EXPLAINS HOW TO OBSERVE THEM.

# OBSERVING
*the* SOLAR SYSTEM

# THE SOLAR SYSTEM

The Solar System is a collection of planets and moons with the Sun at the center. It contains nine planets including Earth, over sixty moons, and numerous other celestial objects together with thousands of asteroids. The Sun contains over 99 per cent of all the matter in the Solar System, and its powerful gravity holds the entire system together. Each object, from the smallest asteroid, to Jupiter, the largest planet, is in orbit around the Sun. At the very edge of the Solar System is a cloud of comets that extends half way to the nearest star.

**Pluto**
*Furthest planet from the Sun, Pluto has an eccentric orbit. It strays from the planetary plane, and once every orbit, its path takes it inside the orbit of Neptune.*

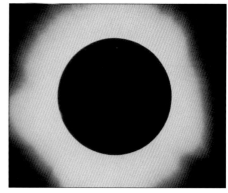

**Saturn**
*The planet Saturn has the most extensive, colorful, and complex ring system, together with the largest family of moons of any planet in the Solar System.*

## PLANETARY MOONS

Most planets in the Solar System have satellites (moons). Only two planets, Mercury and Venus, have none at all. Earth and Pluto have one each, the Moon and Charon, and the other planets have two or more. Moons are minor members of the Solar System but the largest are comparable in size to the smaller planets. The large moons are all circular but the smaller ones, such as Mars's moons Phobos and Deimos, are potato-shaped.

**Titania**
*Uranus's largest moon is about half the size of Earth's Moon. It is pitted with craters.*

**Io**
*One of Jupiter's moons, Io has a constantly changing surface due to volcanic activity.*

**Moon**
*Earth's Moon is the closest celestial object to us and a familiar sight in our sky.*

## ORBITS OF THE PLANETS

The Sun and the planets are shown above. (Sizes and distances are not to scale.) Everything in the Solar System, with the exception of comets, follows an orbit close to the plane of the Sun's equator and moves around the Sun counterclockwise when seen from above the Sun's north pole. Planets and moons usually spin counter-clockwise on their own axes.

## SPECIAL EFFECTS

Spectacular astronomical effects and events can occasionally be seen from Earth. From time to time, for instance, a comet with a brilliant head and tail is clearly visible when it visits our skies (pp.76–7). A solar eclipse is a rare, but memorable, event that occurs only once or twice a year (pp.70–1). Meteor showers happen at regular and predictable times throughout the year (pp.74–5).

**Comet**
*A comet with its glowing head and tail is an unforgettable sight in the sky. The brightest comets can be seen with the naked eye alone.*

**Solar eclipse**
*A solar eclipse occurs when Earth, Moon, and Sun are aligned and the Moon blots out the Sun's disc, allowing the corona to be seen.*

## The Sun
*The Sun, our local star, dominates the Solar System and its gravity holds it all together. We receive light and heat on Earth from the Sun.*

## Mars
*The first planet beyond Earth, Mars rotates on its axis in a similar time to Earth but it takes almost twice as long – 687 days – to make one complete orbit of the Sun.*

## Jupiter
*Jupiter is immense: it contains two and a half times the mass of all the other planets put together. It also spins faster than any other planet. One rotation takes only ten hours.*

## Uranus
*This planet is remote: it is twice as far from the Sun as Saturn. Uranus orbits the Sun, tipped on its side.*

## Neptune
*The more distant a planet is from the Sun, the longer its orbit. Nepune's orbital period is almost 165 years.*

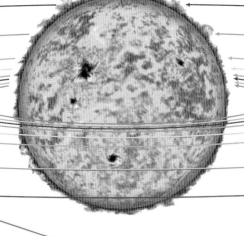

## Venus
*The second closest planet to the Sun, Venus is similar in size to Earth. It spins clockwise, in the opposite direction to other planets.*

## Earth
*Our home planet, Earth, is the only one that has water and supports life. Earth takes 24 hours to spin on its axis and one year, or 365 days, to complete one orbit of the Sun.*

## Mercury
*The closest planet to the Sun, Mercury moves faster than any other planet in the Solar System. It was named after the Roman messenger of the Gods.*

## Meteor
*A meteor is a bright streak of light that can be seen across the sky. It is caused by debris from a comet burning up in Earth's atmosphere.*

## SCALE OF THE PLANETS
Jupiter is the largest planet in the Solar System; it would take only ten Jupiters to fit across the face of the Sun. Saturn, Uranus, and Neptune are also giants. The others are smaller; Earth would fit across the face of Jupiter 11 times. Venus is a little smaller than Earth, then comes Mars, Mercury, and Pluto, the smallest planet.

Mercury    Venus    Earth    Mars

Neptune

Uranus

Jupiter             Saturn              Pluto

# OUR VIEW OF THE SOLAR SYSTEM

From Earth, we can see with the naked eye at least one of every type of object that exists in the Solar System. Even though we are familiar with the spectacular images sent back to Earth by spaceprobes or the Hubble Space Telescope, nothing can equal the thrill of seeing and observing an object such as a planet or comet for oneself. Many objects can be seen with the naked eye, but binoculars or a telescope will show more detail. Once in view, an object can be observed over several nights, or even months, as it travels through the Solar System.

**Looking at the planets**
*All the planets except Neptune and Pluto can be seen with the naked eye if conditions are good, but they are easily mistaken for stars.*

## THE SOLAR SYSTEM AND EARTH

Earth and the other planets all orbit the Sun in much the same plane (p.36). When they are viewed from Earth, the Sun, planets, and the Moon appear to keep close to an imaginary line that circles Earth, although it is really Earth and the planets that are traveling around the Sun. This imaginary line is known as the ecliptic.

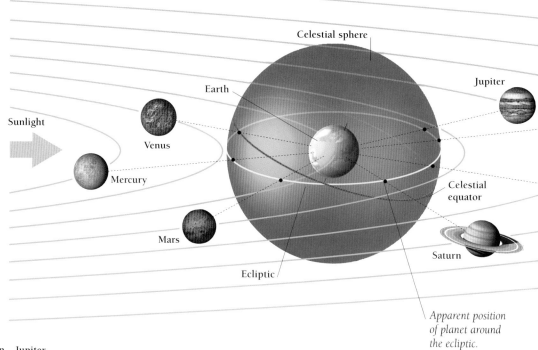

Sunlight
Mercury
Venus
Mars
Earth
Celestial sphere
Jupiter
Celestial equator
Saturn
Ecliptic

*Apparent position of planet around the ecliptic.*

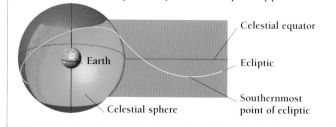

Ecliptic  Venus  Moon  Jupiter

**Viewing the ecliptic from Earth**
*The planets all follow paths that are close to the ecliptic—within a few degrees of it—to its north or south. The position of the ecliptic in the sky depends on your particular location on Earth (pp.14–15).*

## THE ECLIPTIC ON SKY MAPS

The ecliptic crosses the equator of the celestial sphere (p.16) twice, once as the Sun moves from the northern celestial hemisphere and enters the southern, and once as it moves from the southern and into the northern. The band of sky marked by the northernmost point and the southernmost point on the ecliptic is used for the planetary locator maps on pp.40–61.

Earth
Celestial sphere
Celestial equator
Ecliptic
Southernmost point of ecliptic

## RETROGRADE MOTION

The planets seem to move across our sky against the backdrop of stars, in a west to east direction. However, a planet can sometimes appear to move in the opposite direction. This backward movement is an illusion. All planets move forward on their paths around the Sun. The illusion is caused by our viewing position on Earth. When a superior planet—one that is more distant from the Sun than Earth—is overtaken by the faster-moving Earth, the planet's motion becomes retrograde, appearing to move backward temporarily.

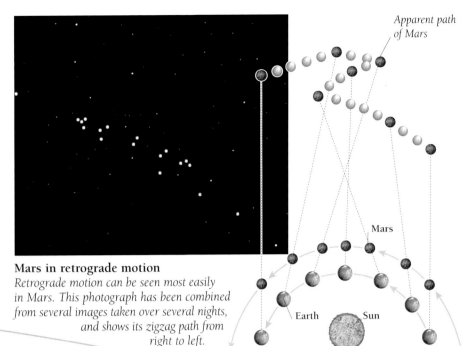

*Apparent path of Mars*

**Mars in retrograde motion**
*Retrograde motion can be seen most easily in Mars. This photograph has been combined from several images taken over several nights, and shows its zigzag path from right to left.*

**Why retrograde motion happens**
*Earth, on the inside track, overtakes Mars as both planets travel in orbit around the Sun. Mars appears to move backward in Earth's sky.*

Uranus

*Actual position of the planet and its orbit*

Neptune

Jupiter    Saturn

## WHEN TO OBSERVE THE PLANETS

The planets Mercury and Venus orbit the Sun within Earth's orbit, and are described as inferior planets. When we look toward them, we are looking toward the Sun. The best time to see an inferior planet is at elongation (p.47). Planets with orbits beyond Earth's are described as superior. Superior planets, unlike inferior ones, can be observed away from the Sun and are best seen at opposition—when they are directly opposite the Sun.

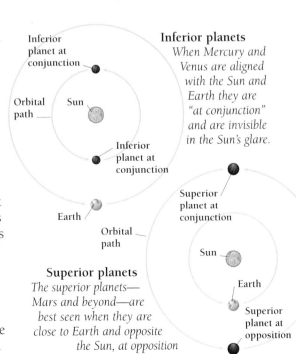

Inferior planet at conjunction

Orbital path

Sun

Inferior planet at conjunction

Earth

Orbital path

**Superior planets**
*The superior planets—Mars and beyond—are best seen when they are close to Earth and opposite the Sun, at opposition*

Superior planet at conjunction

Sun

Earth

Superior planet at opposition

**Inferior planets**
*When Mercury and Venus are aligned with the Sun and Earth they are "at conjunction" and are invisible in the Sun's glare.*

**Planets and the Moon**
*Because the planets and the Moon move across the same band of sky, two or more may sometimes appear to be positioned close together. If two actually become aligned along the ecliptic, they are said to be "in conjunction."*

# MERCURY

Mercury is a dark, gray world. Even though it is the nearest planet to the Sun, it reflects only 6 percent of the sunlight that falls on it. The planet moves quickly, taking just under 88 Earth days to orbit the Sun. Because it keeps so close to the Sun, Mercury can never be seen in a fully dark sky, and its surface is impossible to observe from Earth.

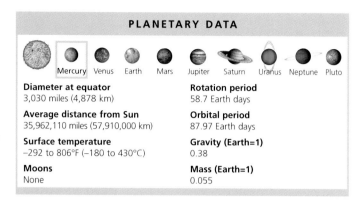

### PLANETARY DATA

Mercury  Venus  Earth  Mars  Jupiter  Saturn  Uranus  Neptune  Pluto

**Diameter at equator**
3,030 miles (4,878 km)

**Average distance from Sun**
35,962,110 miles (57,910,000 km)

**Surface temperature**
−292 to 806°F (−180 to 430°C)

**Moons**
None

**Rotation period**
58.7 Earth days

**Orbital period**
87.97 Earth days

**Gravity (Earth=1)**
0.38

**Mass (Earth=1)**
0.055

## LOCATING MERCURY 1999–2010

To find Mercury, use the map (below) to discover which constellation the planet is in, then use the planisphere to find the exact location of the constellation for your latitude and the time. All the positions marked on the map are for times when Mercury is at its greatest distance from the Sun in Earth's sky, i.e. at elongation (p.47), and therefore in the best position for observing from Earth.

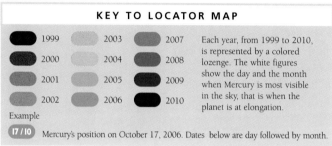

### KEY TO LOCATOR MAP

| | | |
|---|---|---|
| 1999 | 2003 | 2007 |
| 2000 | 2004 | 2008 |
| 2001 | 2005 | 2009 |
| 2002 | 2006 | 2010 |

Example

**17 / 10**  Mercury's position on October 17, 2006. Dates below are day followed by month.

Each year, from 1999 to 2010, is represented by a colored lozenge. The white figures show the day and the month when Mercury is most visible in the sky, that is when the planet is at elongation.

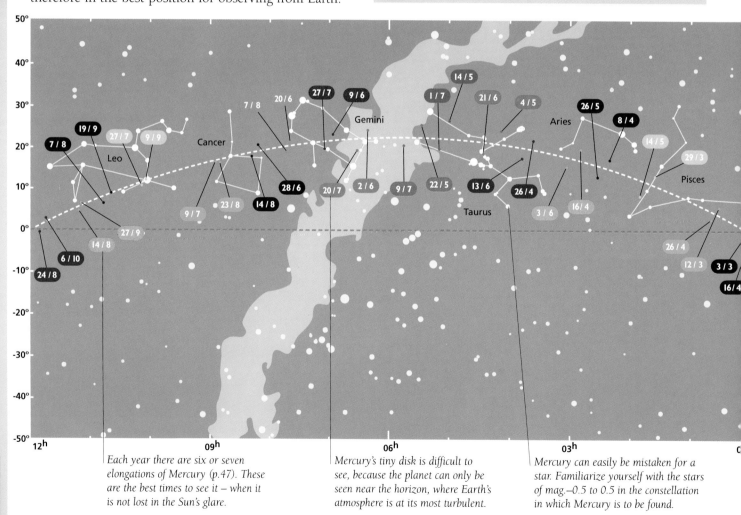

*Each year there are six or seven elongations of Mercury (p.47). These are the best times to see it – when it is not lost in the Sun's glare.*

*Mercury's tiny disk is difficult to see, because the planet can only be seen near the horizon, where Earth's atmosphere is at its most turbulent.*

*Mercury can easily be mistaken for a star. Familiarize yourself with the stars of mag.−0.5 to 0.5 in the constellation in which Mercury is to be found.*

MERCURY

MERCURY

VENUS

MARS

JUPITER

SATURN

URANUS

NEPTUNE

PLUTO

MOON

SUN

## MERCURY'S SURFACE

In the 1970s, Mariner 10 flew by Mercury and sent back pictures of a rocky planet which, like the Moon, is pitted with thousands of craters. Mercury also has long, low ridges called scarps. The craters were formed by a bombardment of space rocks, and the scarps are the result of surface wrinkling caused by cooling and shrinking. Mercury's surface has remained unchanged for billions of years, as there is no atmosphere to alter its features.

*Ray craters are similar to those on the Moon.*

**Scale**
*Mercury has a diameter of 3,030 miles (4,878 km), which is less than half the diameter of Earth.*

*Craters are generally circular, with rim deposits; the larger ones have inner walls and peaks.*

*The Caloris Basin is the largest feature on Mercury. It was caused by a rock crashing into the planet.*

*Scarps are up to 1.8 miles (3 km) high and 300 miles (500 km) long.*

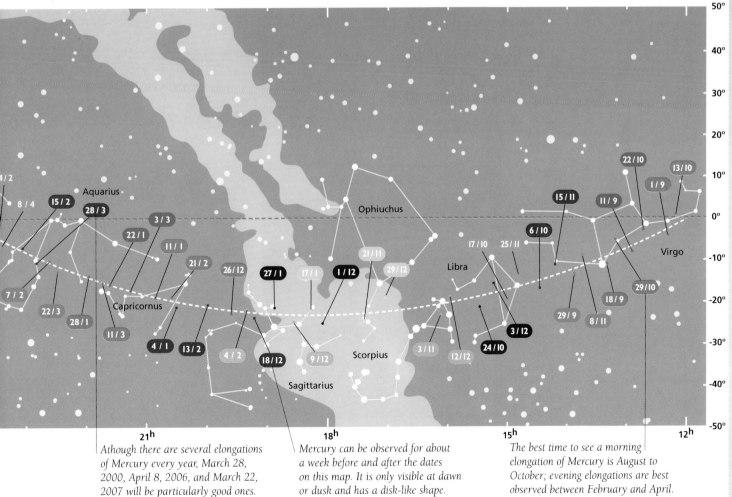

*Athough there are several elongations of Mercury every year, March 28, 2000, April 8, 2006, and March 22, 2007 will be particularly good ones.*

*Mercury can be observed for about a week before and after the dates on this map. It is only visible at dawn or dusk and has a disk-like shape.*

*The best time to see a morning elongation of Mercury is August to October; evening elongations are best observed between February and April.*

# OBSERVING MERCURY

Every year, there are six occasions when Mercury can be seen at its best; three times to the east of the Sun, and three times to the west. On these occasions Mercury is found close to the horizon in a twilight sky, either in the morning or evening. When it is west of the Sun, it appears just before sunrise; when to the east, it appears the hour or so after sunset. Mercury can then be observed for about two weeks before it again goes out of view, hidden by the Sun's light. Every few years a transit of Mercury gives observers the chance to make an unusual observation.

*Seen from Earth, Mercury appears to begin crossing the Sun.*

*Mercury has completed half of its transit.*

Earth

Mercury

Line of sight

*Mercury has crossed to the far edge of the Sun's disk and its transit is complete.*

## LOOKING AT MERCURY

**👁 Naked-eye view**
*At its brightest, Mercury can outshine Sirius (p.113). Mercury is always near the horizon, appearing as a flickering point of light due to turbulence in Earth's atmosphere.*

**🔭 Binocular view**
*There is no advantage to using binoculars to view Mercury, but they may help you locate it. Make sure the Sun is below the horizon, and sweep the binoculars back and forth.*

**🔭 Telescope view**
*Even through a small telescope, Mercury looks like a twinkling star-like object. Larger telescopes, however, show its shape and its phase should also be discernible.*

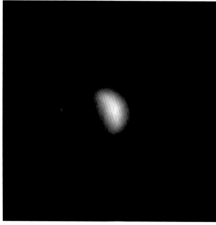

**💻 CCD image view**
*The phases become clear, although no surface features can be seen. A full phase is never observed because, when it is full, Mercury is out of sight on the far side of the Sun.*

## TRANSIT OF MERCURY

Mercury's orbit usually takes it above or below the Sun's disk in the sky, when seen from Earth. When the Sun, Mercury, and Earth are aligned, Mercury appears to cross over the face of the Sun. This phenomenon is known as a transit of Mercury.

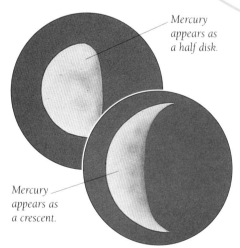
*Mercury appears as a half disk.*

*Mercury appears as a crescent.*

**Mercury's phases**
*Mercury has phases like Venus and is best seen at elongation (p.47), when it is a half disk. Observers with high-powered telescopes have seen light and dark regions on Mercury, although these do not bear any relation to surface features that have been identified.*

MERCURY

VENUS

MARS

JUPITER

SATURN

URANUS

NEPTUNE

PLUTO

MOON

SUN

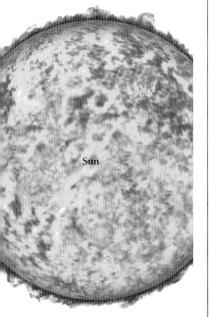

*Mercury's orbit around the Sun.*

Sun

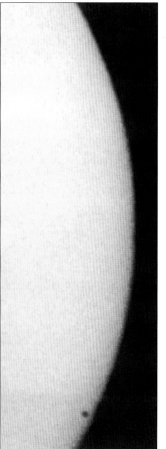
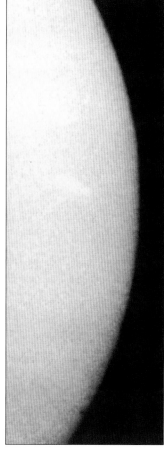

## 1973 TRANSIT

These three images show the final stages of Mercury's transit across the Sun in 1973. Mercury is a tiny black dot against the huge face of the Sun. The planet takes several hours to cross the Sun's disk before it disappears off the right-hand edge. The duration of a transit depends on where Mercury is seen to be crossing the Sun. A transit of the Sun's equator lasts longest.

**Mercury in the evening twilight sky**
*Mercury is the white dot in the center of the picture. The Sun has just set, and Mercury will soon also disappear from view below the horizon. The planet's proximity to the horizon presents a major observing problem, because the turbulent air at the horizon makes it extremely difficult to obtain a steady and satisfactory view of the planet.*

### BEST TIMES TO OBSERVE

Mercury is best observed when it is at elongation (p.47). At its greatest eastern elongation (east of the Sun) the planet can be seen in the western sky after sunset. When at greatest western elongation (west of the Sun), it is in the east just before sunrise. The locator map on pp.40–41 shows the elongations from 1999 to 2010.

### DATES OF TRANSITS

The duration of a transit of the Sun depends on Mercury's position and the planet's path across the Sun's disk.

Nov 15, 1999—at 21.43 for 52 mins
May 7, 2003—at 7.53 for 5 hrs 18 mins
Nov 8, 2006—at 21.42 for 4 hrs 58 mins
May 9, 2016—at 15.00 for 7 hrs 30 mins

# VENUS

Venus is the second planet from the Sun and its orbit brings it within 26 million miles (42 million km) of Earth, closer than any other planet. Venus is the most brilliant object in the night sky after the Moon. The bright surface we see is the top layer of a dense carbon-dioxide atmosphere. Hidden from view is a hot, inhospitable, volcanic world.

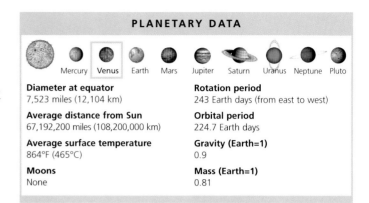

### PLANETARY DATA

Mercury | Venus | Earth | Mars | Jupiter | Saturn | Uranus | Neptune | Pluto

| | |
|---|---|
| **Diameter at equator**<br>7,523 miles (12,104 km) | **Rotation period**<br>243 Earth days (from east to west) |
| **Average distance from Sun**<br>67,192,200 miles (108,200,000 km) | **Orbital period**<br>224.7 Earth days |
| **Average surface temperature**<br>864°F (465°C) | **Gravity (Earth=1)**<br>0.9 |
| **Moons**<br>None | **Mass (Earth=1)**<br>0.81 |

## LOCATING VENUS 1999–2010

To find Venus, use the map (below) to discover which constellation the planet is in, then use the planisphere to find the exact location of the constellation for your latitude and the time. You should be able to see Venus with the naked eye in the early morning or early evening. The planet keeps close to the line of the ecliptic and, about every 19 months, it goes into retrograde motion.

### KEY TO LOCATOR MAP

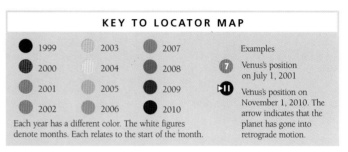

| | | | |
|---|---|---|---|
| ● 1999 | 2003 | 2007 | Examples |
| 2000 | 2004 | 2008 | Venus's position on July 1, 2001 |
| 2001 | 2005 | 2009 | Venus's position on November 1, 2010. The arrow indicates that the planet has gone into retrograde motion. |
| 2002 | 2006 | 2010 | |

Each year has a different color. The white figures denote months. Each relates to the start of the month.

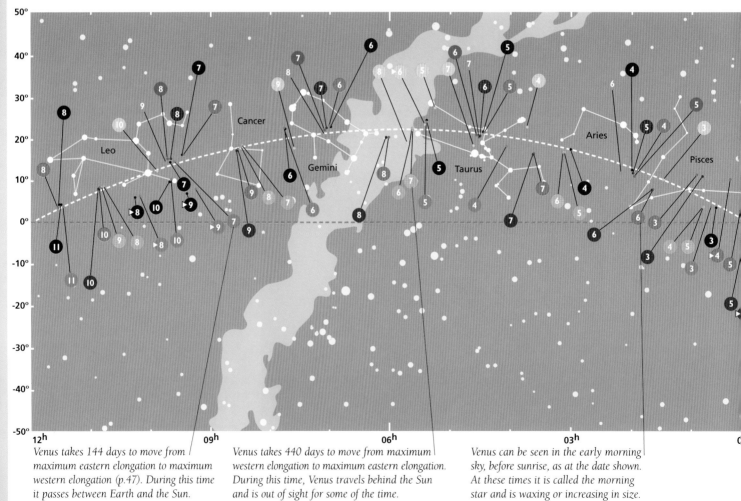

*Venus takes 144 days to move from maximum eastern elongation to maximum western elongation (p.47). During this time it passes between Earth and the Sun.*

*Venus takes 440 days to move from maximum western elongation to maximum eastern elongation. During this time, Venus travels behind the Sun and is out of sight for some of the time.*

*Venus can be seen in the early morning sky, before sunrise, as at the date shown. At these times it is called the morning star and is waxing or increasing in size.*

## VENUS'S SURFACE

One of the most successful probes to Venus, Magellan, used radar to map almost the entire planet between 1989 and 1992. It became apparent that Venus is a world of highlands, depressions and rolling plains, covered by lava flows, impact craters, and canyons. More than three quarters of the planet has been affected by volcanic eruptions and many volcanoes may still be active. This picture shows how Venus looks, far below the clouds.

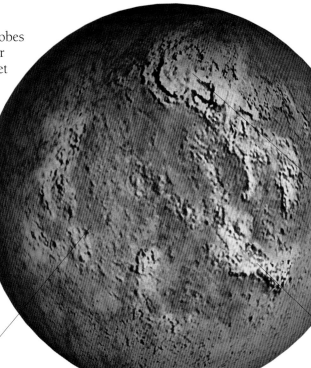

**Scale**
*Venus has a diameter of about 7,500 miles (12,100 km) which is a little less than that of Earth.*

*Ishtar Terra is one of the two main highland areas on Venus. It includes the Maxwell Montes, the highest mountain range.*

*Rolling plains, formed from volcanic lava, cover the entire planet.*

*Aphrodite Terra lies near the equator and is the largest upland region on Venus.*

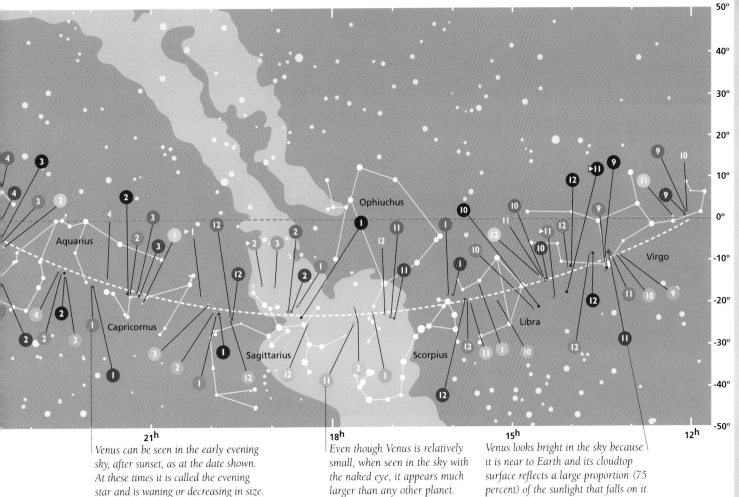

*Venus can be seen in the early evening sky, after sunset, as at the date shown. At these times it is called the evening star and is waning or decreasing in size.*

*Even though Venus is relatively small, when seen in the sky with the naked eye, it appears much larger than any other planet.*

*Venus looks bright in the sky because it is near to Earth and its cloudtop surface reflects a large proportion (75 percent) of the sunlight that falls on it.*

45

SUN MOON PLUTO NEPTUNE URANUS SATURN JUPITER MARS VENUS MERCURY

# OBSERVING VENUS

Venus, like Mercury, never moves far from the Sun. Unlike Mercury, however, Venus can be seen in a dark sky, because it is further away from the Sun. We see it shining brightly in the morning or evening. The planet's maximum magnitude is –4.7, but its brightness varies, as does its apparent size in the sky. As it moves closer to Earth, the planet's disk appears to grow larger in size, but because Venus also undergoes a phase cycle, we actually see less of it as it approaches Earth during its orbit, and more of it as it recedes from Earth.

## VENUS' PHASES

Venus goes through a phase cycle similar to the Moon's (pp.62–3), as only one side of it is lit by the Sun. We see differing amounts of the sunlit side. The pictures below cover about a third of Venus' phase cycle, showing how the planet's appearance changes from gibbous, or three-quarters full (left), to a large disk with a tiny band of light when Venus lies directly between the Sun and Earth (far right).

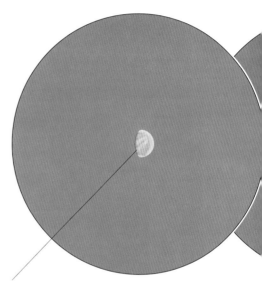

*We see the maximum amount of Venus' disk when the planet is at its furthest visible point from Earth, although it appears small due to its great distance.*

## LOOKING AT VENUS

### 🌑 Naked-eye view
*Venus looks like a brilliant star in the night sky. It is easily spotted with the naked eye but its light can be so dazzling that this can sometimes be a barrier to seeing it clearly.*

### 🔭 Binocular view
*The phases become apparent when Venus is seen with binoculars. Venus' cycle can be followed as the sunlit part of the disk increases and decreases, and the planet alters in size.*

### 🔭 Telescope view
*The disk of Venus appears larger through a telescope, and its phase becomes slightly clearer; otherwise a telescope view is no real improvement on a binocular view.*

### 📷 CCD image view
*Venus' phase becomes apparent when seen as a CCD image. The planet appears as a yellowish-white disk. Faint markings can also be detected in the upper cloud layer.*

---

### BEST TIMES TO OBSERVE

The best time to see Venus is when it is at its greatest elongation east or west of the Sun (see below). When at its greatest elongation east, it will be seen in the evening sky; when it is at its greatest elongation west, it will appear in the early morning sky, before sunrise.

### DATES OF ELONGATIONS

| | |
|---|---|
| E Jun 11,1999 | E Nov 3, 2005 |
| W Oct 31, 1999 | W Mar 25, 2006 |
| E Jan 17, 2001 | E Jun 9, 2007 |
| W Jun 8, 2001 | W Oct 28, 2007 |
| E Aug 22, 2002 | E Jan 14, 2009 |
| W Jan 11, 2003 | W Jun 5, 2009 |
| E Mar 29, 2004 | E Aug 20, 2010 |
| W Aug 17, 2004 | |

**Venus and the Moon in the evening sky**
*Once the Sun has set below the horizon, Venus can be seen shining like a brilliant star in the early evening sky; it is only outshone by the Moon, and will soon disappear below the horizon. When the planet is seen, as here, in the evening sky, it is the eastern side of its face that is illuminated. When it rises in the morning sky, it is the western side of the planet that is lit up by the Sun.*

*At its nearest point to Earth, Venus is large, bright, and easy to see. Light and dark areas can be seen on its surface.*

*As Venus moves closer to Earth the planet appears to get larger but less of its surface is visible.*

*Venus is in its crescent phase. The tips (cusp caps) of the crescent are noticeably brighter than the rest of the crescent.*

*Venus is almost directly between the Sun and the Earth, and only a thin sliver of light extends almost around the entire planet.*

## TRANSIT OF VENUS, 1882

Venus, like Mercury, makes transits of the Sun (pp.42–3). Transits of Venus are rare, however, and occur at dates that fall relatively close together. The last pair was in 1874 and 1882 and the next pair will be in 2004 and 2012.

## ELONGATION OF INFERIOR PLANETS

Like Mercury (pp.40–3), Venus is an inferior planet lying between Earth and the Sun. The ease with which we see it depends on its position relative to Earth and the Sun. When it is between Earth and the Sun, it is lost in the Sun's glare. When it is behind the Sun it is out of sight. The best times for viewing are when the planet is at an angle, east or west of the Sun. This is the angle of elongation, and the planet is said to be "at elongation."

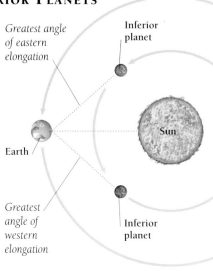

*Greatest angle of eastern elongation*

*Inferior planet*

*Sun*

*Earth*

*Greatest angle of western elongation*

*Inferior planet*

# MARS

Mars is a prominent red object in the night sky. It is only half Earth's size, and some 48.5 million miles (78 million km) away, but it is easy to see for much of the year. Mars is the only planet with landscape features that can be seen from Earth, and is the planet that most resembles Earth. It has polar ice caps, seasons, and a surface that once had running water.

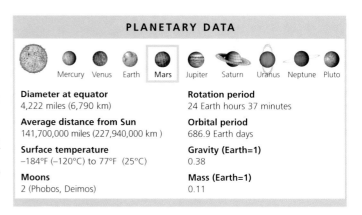

**PLANETARY DATA**

Mercury  Venus  Earth  Mars  Jupiter  Saturn  Uranus  Neptune  Pluto

**Diameter at equator**
4,222 miles (6,790 km)

**Rotation period**
24 Earth hours 37 minutes

**Average distance from Sun**
141,700,000 miles (227,940,000 km )

**Orbital period**
686.9 Earth days

**Surface temperature**
−184°F (−120°C) to 77°F (25°C)

**Gravity (Earth=1)**
0.38

**Moons**
2 (Phobos, Deimos)

**Mass (Earth=1)**
0.11

## LOCATING MARS 1999–2010

To find Mars, use the map below to discover which constellation it is in. Then use the planisphere to find the location of the constellation for your latitude and the time. If viewing conditions are good, you should be able to see Mars with the naked eye. Mars keeps close to the ecliptic and although it travels west to east in the sky (right to left on the map), it regularly goes into retrograde motion (p.39).

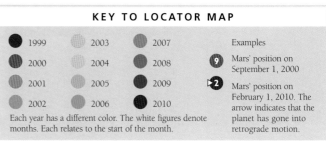

**KEY TO LOCATOR MAP**

| | | | Examples |
|---|---|---|---|
| 1999 | 2003 | 2007 | |
| 2000 | 2004 | 2008 | Mars' position on September 1, 2000 |
| 2001 | 2005 | 2009 | Mars' position on February 1, 2010. The arrow indicates that the planet has gone into retrograde motion. |
| 2002 | 2006 | 2010 | |

Each year has a different color. The white figures denote months. Each relates to the start of the month.

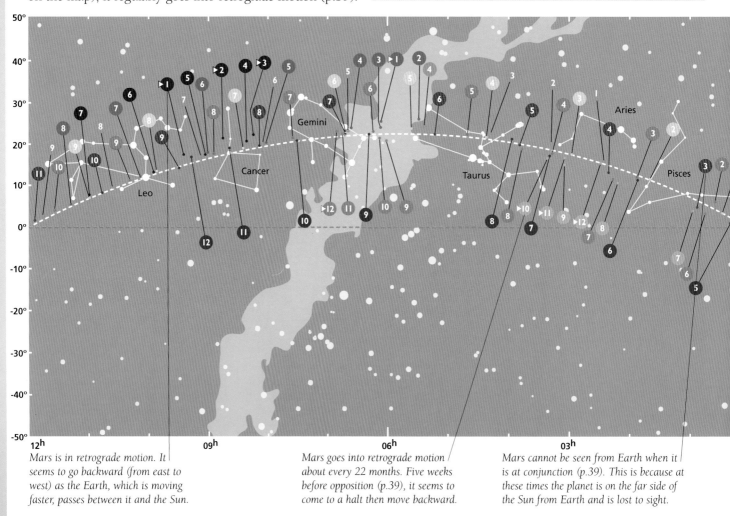

*Mars is in retrograde motion. It seems to go backward (from east to west) as the Earth, which is moving faster, passes between it and the Sun.*

*Mars goes into retrograde motion about every 22 months. Five weeks before opposition (p.39), it seems to come to a halt then move backward.*

*Mars cannot be seen from Earth when it is at conjunction (p.39). This is because at these times the planet is on the far side of the Sun from Earth and is lost to sight.*

## MARS'S SURFACE

Mars is a small rocky planet that is noticeably red in color. The color comes from the red iron-oxide rock and dust that cover most of the surface. Fierce storms blow dust around, which shows up as light and dark patches when seen from Earth. Vast, extinct volcanoes rise above the plains, and a giant canyon system splits the planet across the middle. The surface is criss-crossed by narrow channels formed long ago by running water.

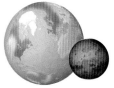

**Scale**
*With a diameter of 4,222 miles (6,790 km), Mars is just over half the size of Earth.*

*Dry valleys, formed in the past by running water, meander for hundreds of miles.*

*Valles Marineris is a canyon system caused by the faulting and collapse of the Martian surface. It is more than 2,500 miles (4,000 km) long.*

*Olympus Mons is 16 miles (26 km) high and is the largest volcano in the Solar System.*

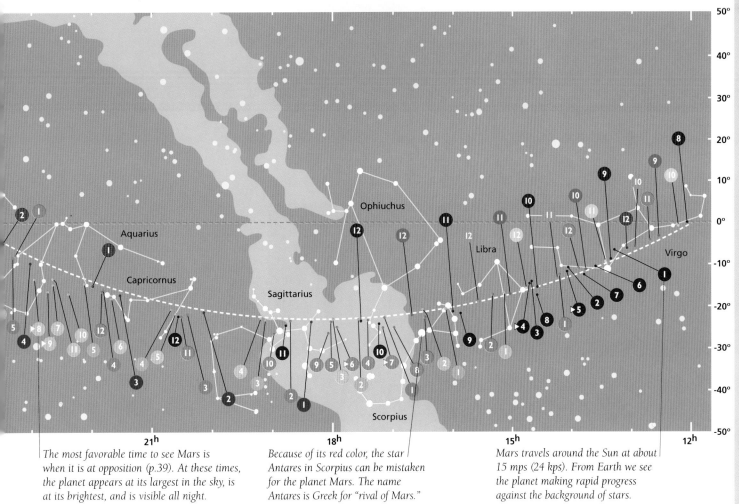

Aquarius · Capricornus · Sagittarius · Scorpius · Ophiuchus · Libra · Virgo

21ʰ · 18ʰ · 15ʰ · 12ʰ

*The most favorable time to see Mars is when it is at opposition (p.39). At these times, the planet appears at its largest in the sky, is at its brightest, and is visible all night.*

*Because of its red color, the star Antares in Scorpius can be mistaken for the planet Mars. The name Antares is Greek for "rival of Mars."*

*Mars travels around the Sun at about 15 mps (24 kps). From Earth we see the planet making rapid progress against the background of stars.*

# OBSERVING MARS

Like all the planets, Mars has no light of its own but shines through reflected sunlight. Mars is easier to see than the inferior planets, Mercury and Venus, because it lies beyond the orbit of Earth and we do not look toward the Sun when we observe it. The most favorable time to see Mars is at opposition (p.39), when it is close to Earth and high in the sky. Because of the eccentricity of the orbit of Mars, this occurs only about once every two years, and then the planet comes within 36 million miles (59 million km) of Earth.

## LOOKING AT MARS

**Naked-eye view**
*Mars is visible to anyone with good eyesight. It is only outshone by Venus at its brightest. In the picture above, Mars (left) is in conjunction with Jupiter (right).*

**Binocular view**
*Through binoculars, it becomes obvious that Mars is a planet, because of its disk-like shape. If the binoculars are powerful enough, the Martian polar ice caps can just be seen.*

**Telescope view (inverted)**
*Seen through a telescope, Mars is noticeably red, and its surface features will be brought into view. The polar ice caps and the dark equatorial regions can be observed.*

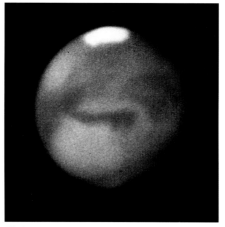

**CCD image view (inverted)**
*The south pole (top) is tilted toward Earth and can be seen as a white patch. Shaded markings are now clearly visible. The dark patch on the left-hand edge is Syrtis Major.*

**9.29 p.m.** *Landscape features on the Martian surface can be clearly seen.*

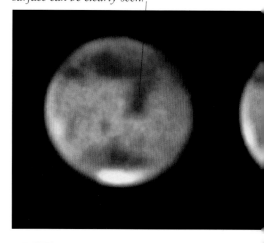

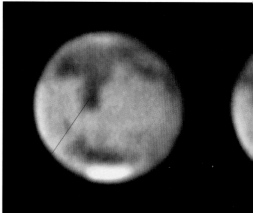

**11.23 p.m.** *Syrtis Major moves to far left*

## THE ROTATION OF MARS

Mars turns once every 24 hours 37 minutes. Part of its rotation can be observed in an evening. The inverted images above are through a telescope.

### BEST TIMES TO OBSERVE

When Mars is at opposition (p.39) it is at its most favorable position for observation. This is because Earth lies directly between the Sun and Mars, and the planet is above the horizon all night long. The opposition in August 2003 will be particularly favorable.

### DATES OF OPPOSITIONS

| | |
|---|---|
| Apr 24, 1999 | Nov 7, 2005 |
| Jun 13, 2001 | Dec 24, 2007 |
| Aug 28, 2003 | Jan 29, 2010 |

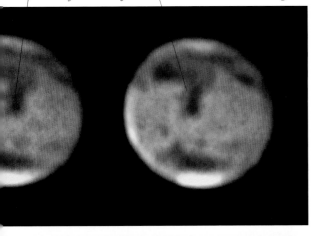

**10.05 p.m.** *Syrtis Major has moved closer to the center of Mars' surface.*

**10.50 p.m.** *Surface features are in the same position about 37 minutes later each night.*

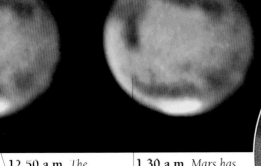

**12.50 a.m.** *The north polar ice cap is tilted toward Earth.*

**1.30 a.m.** *Mars has completed about one sixth of its rotation.*

## THE MOONS OF MARS

The two Martian moons, Phobos and Deimos, can be observed from Earth. They are difficult to see because Mars can easily outshine them. Any observer lucky enough to spot them will only see tiny specks of light. We only know what the moons look like through spaceprobes.

Mars

Phobos' orbit

Deimos is 12,400 miles (20,065 km) from Mars

Phobos is 3,700 miles (5,980 km) from Mars

Phobos

Deimos' orbit

Deimos

### Deimos
*Only 9 miles (16 km) in length, Deimos is the smaller of the two Martian moons. It shines at about mag.13 and orbits Mars every 30 hours.*

### Phobos
*This moon is approximately 15 miles (28 km) long. It orbits Mars in less than eight hours. Like Deimos, Phobos is thought to be an asteroid that was captured by Mars' gravity.*

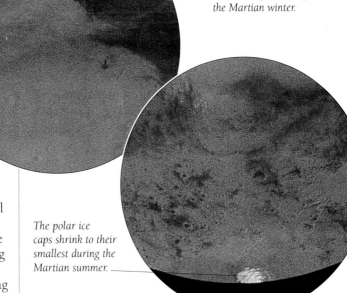

*The polar ice caps change in size and are at their largest during the Martian winter.*

*The polar ice caps shrink to their smallest during the Martian summer.*

## OLYMPUS MONS

Thanks to pictures sent back by spaceprobes, we now have detailed knowledge of what the surface of Mars is like. Landing craft have produced colorful, sharp images of Mars' red, rocky, desert terrain that is impossible to see from Earth, even with powerful telescopes. Spaceprobes orbiting the planet show gaping canyons and huge volcanoes, including the vast Olympus Mons (left), which is only visible from Earth as a light patch.

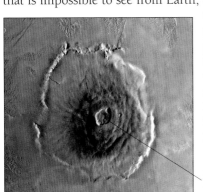

*The crater of Olympus Mons is 53 miles (5 km) wide.*

## POLAR ICE CAPS

The easiest details to pick out on the Martian surface are the polar ice caps which are Mars' equivalent of the Arctic and Antarctic regions on Earth. A 3-in (80-mm) telescope will bring out these polar caps along with dark markings near the equatorial regions. For more detailed observations, a more powerful telescope is needed.

# JUPITER

Jupiter is a giant planet lying far beyond the asteroid belt (pp.78–9). It is composed of gas and liquid with a solid core and is larger than all the other planets put together. Even though Jupiter is so remote, bright sunlight reflecting off its thick atmosphere makes it visible from Earth for about ten months of the year. Its colorful and fast-moving bands of clouds are fascinating to watch through a telescope.

## LOCATING JUPITER 1999–2010

To find Jupiter, use the map (below) to discover which constellation the planet is in. Then use the planisphere to find the location of the constellation for your latitude and the time. Jupiter keeps close to the ecliptic. It is easy to spot with the naked eye (only the Moon and Venus are brighter) and regularly goes into retrograde motion.

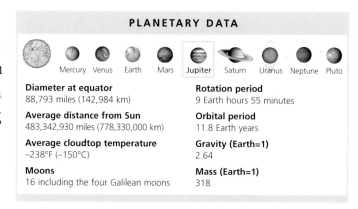

### PLANETARY DATA

Mercury  Venus  Earth  Mars  Jupiter  Saturn  Uranus  Neptune  Pluto

**Diameter at equator**
88,793 miles (142,984 km)

**Average distance from Sun**
483,342,930 miles (778,330,000 km)

**Average cloudtop temperature**
−238°F (−150°C)

**Moons**
16 including the four Galilean moons

**Rotation period**
9 Earth hours 55 minutes

**Orbital period**
11.8 Earth years

**Gravity (Earth=1)**
2.64

**Mass (Earth=1)**
318

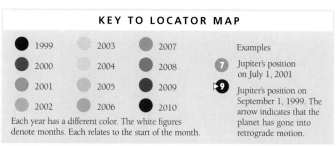

### KEY TO LOCATOR MAP

| | | | |
|---|---|---|---|
| ● 1999 | 2003 | 2007 | Examples |
| ● 2000 | 2004 | 2008 | ⑦ Jupiter's position on July 1, 2001 |
| ● 2001 | 2005 | ● 2009 | ▷⑨ Jupiter's position on September 1, 1999. The arrow indicates that the planet has gone into retrograde motion. |
| 2002 | 2006 | ● 2010 | |

Each year has a different color. The white figures denote months. Each relates to the start of the month.

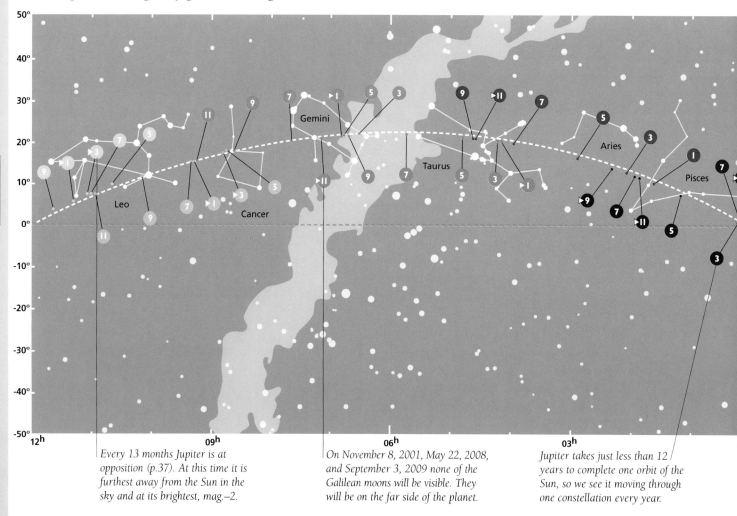

*Every 13 months Jupiter is at opposition (p.37). At this time it is furthest away from the Sun in the sky and at its brightest, mag.–2.*

*On November 8, 2001, May 22, 2008, and September 3, 2009 none of the Galilean moons will be visible. They will be on the far side of the planet.*

*Jupiter takes just less than 12 years to complete one orbit of the Sun, so we see it moving through one constellation every year.*

## JUPITER'S SURFACE

The surface we see from Earth consists of clouds of gas, predominantly hydrogen, some helium, and traces of methane and ammonia. Due to the planet's fast spin, the clouds form into bands of light and dark known as zones and belts, respectively.

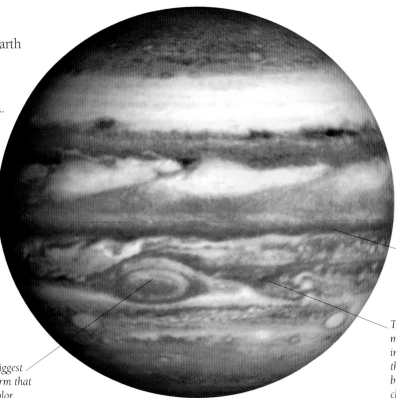

**Scale**
*Jupiter is huge: 11 planets the size of Earth would fit across its face.*

*In 1979, spaceprobe Voyager 1 discovered a ring system that surrounds the planet.*

*The Great Red Spot is Jupiter's biggest surface feature. It is a raging storm that constantly changes in size and color.*

*Adjacent belts and zones have winds blowing in opposite directions, which cause eddies and storms.*

*The bright and dark colors may be the result of alterations in temperature and pressure; the colors change from blue to brown to white, then red as the clouds get higher and cooler.*

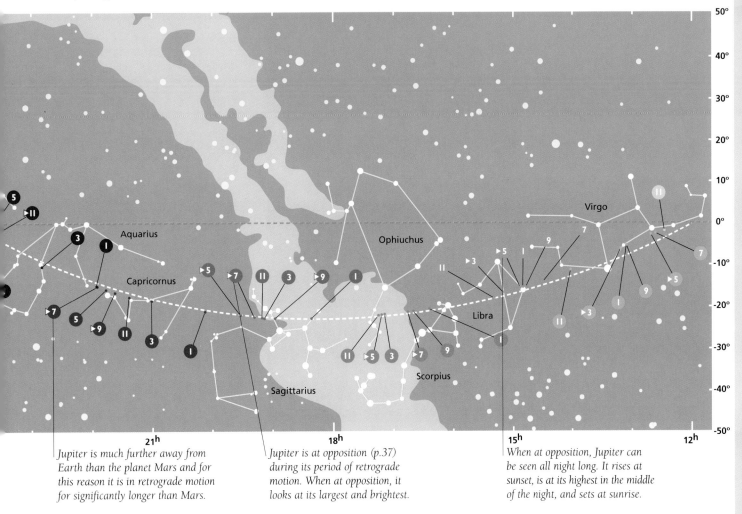

*Jupiter is much further away from Earth than the planet Mars and for this reason it is in retrograde motion for significantly longer than Mars.*

*Jupiter is at opposition (p.37) during its period of retrograde motion. When at opposition, it looks at its largest and brightest.*

*When at opposition, Jupiter can be seen all night long. It rises at sunset, is at its highest in the middle of the night, and sets at sunrise.*

# OBSERVING JUPITER

Jupiter is an interesting planet for all observers. Its brilliant silvery light and disk-like shape make it easy to spot with the naked eye. The planet makes a good binocular subject, and there is always something new to observe with a telescope because our view of the planet's surface is constantly changing due to its rapid rotation. Jupiter has many moons but the four largest, named the Galilean moons after the Italian astronomer Galileo Galilei, shine as brightly as many stars and can be observed with binoculars as they orbit the planet.

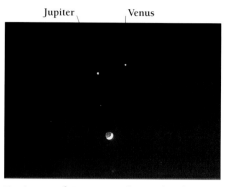

**Jupiter and Venus in the night sky**
*The two brightest planets, Jupiter and Venus, are seen here with Earth's Moon.*

## LOOKING AT JUPITER

**Naked-eye view**
*Jupiter looks brilliant in the night sky. It is easily spotted with the naked eye alone and its disk-like shape is clearly visible. Jupiter is only outshone by the Moon and Venus.*

**Binocular view**
*Through binoculars, Jupiter's Galilean moons can be seen clearly as tiny dots of light. They lie along each side of the line of Jupiter's equator and their positions change over time.*

**Telescope view**
*The belts and zones on the surface of Jupiter are now apparent and the well-known red-brown and yellow-ochre colors can be seen. Surface changes can be observed.*

**CCD image view**
*Jupiter's Individual storm systems and in particular the Great Red Spot can be seen. The bulging equator that gives the planet its slightly oval shape is clearly visible.*

## JUPITER'S MOONS

Jupiter's four largest moons, the Galilean moons, orbit the planet at varying speeds. Their positions change regularly over four nights' observation, as illustrated here. Sometimes they disappear from view altogether as they move behind the planet.

**Io**
*Io is the closest Galilean moon to Jupiter. It orbits Jupiter in less than 2 days.*

**Callisto**
*The most distant moon from Jupiter, Callisto orbits it in just under 17 days.*

**Ganymede**
*Bigger than Mercury, this is the largest moon in the Solar System. It has a complex surface.*

**Europa**
*This moon is smaller than Earth's Moon and has a smooth, icy surface.*

## THE GREAT RED SPOT

This is Jupiter's most noticeable feature, and it was first observed over 300 years ago. The area is an enormous raging storm that changes in size and color, over a period of decades. The Great Red Spot is larger than Earth, reaching, at its largest, about three times Earth's diameter. The color ranges from gray-pink to dark red and this is thought to be the result of phosphorus being brought up from the lower atmosphere.

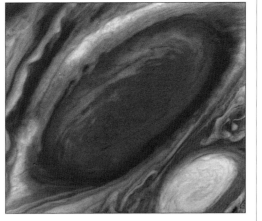

**Spaceprobe picture of the Great Red Spot**

## THE ROTATION OF JUPITER

Jupiter spins more rapidly than any other planet; it completes one rotation in just under ten hours. It does not spin as a solid body, however. The equatorial regions rotate quickest, in 9 hrs 50 mins, and the north and south regions, including the Great Red Spot, take another five minutes to rotate.

*The Great Red Spot takes about two and a half hours to move from the edge to the center of the disk.*

*Jupiter spins from left to right, but in this inverted telescope image it appears to spin from right to left.*

*The black spots are shadows created by Jupiter's Galilean moons.*

*The Great Red Spot is now in the center of the disk.*

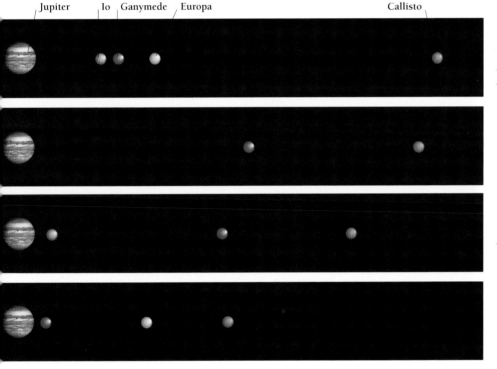

Jupiter    Io   Ganymede   Europa        Callisto

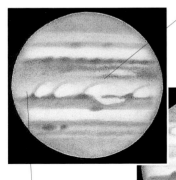

*The widths and colors of the bands can be recorded.*

*Start drawing on the west edge where features disappear first from view.*

*The Great Red Spot is at the top because the planet is viewed through a telescope.*

### Keeping a record

*Drawing will enable you to keep a record of your observations and improve your knowledge of the planet. Observe for about 15 minutes before starting to draw. Then draw for about another 15 minutes: the features will change after this time because of Jupiter's swift rotation.*

### BEST TIMES TO OBSERVE

The best time to observe Jupiter is when it is at opposition (p.39). At this time the Sun shines fully on it, and the planet is at its brightest. The dates below are for when Jupiter is at opposition.

### JUPITER AT OPPOSITION

| | |
|---|---|
| Oct 23, 1999 | May 4, 2006 |
| Nov 28, 2000 | Jun 5, 2007 |
| Jan 1, 2002 | Jul 9, 2008 |
| Feb 2, 2003 | Aug 14, 2009 |
| Mar 4, 2004 | Sep 21, 2010 |
| Apr 3, 2005 | |

# SATURN

Saturn is the sixth planet from the Sun and, like Jupiter, it is a gas giant. Although it is almost twice as remote from Earth as Jupiter, its immense size makes it visible with the naked eye for about 10 months of the year. Saturn has an impressive and colorful ring system and the largest family of moons of any planet in the Solar System.

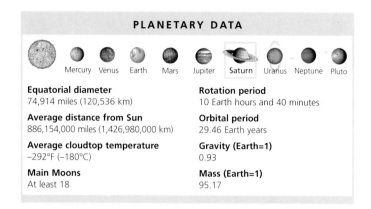

## PLANETARY DATA

Mercury  Venus  Earth  Mars  Jupiter  Saturn  Uranus  Neptune  Pluto

**Equatorial diameter**
74,914 miles (120,536 km)

**Average distance from Sun**
886,154,000 miles (1,426,980,000 km)

**Average cloudtop temperature**
−292°F (−180°C)

**Main Moons**
At least 18

**Rotation period**
10 Earth hours and 40 minutes

**Orbital period**
29.46 Earth years

**Gravity (Earth=1)**
0.93

**Mass (Earth=1)**
95.17

## LOCATING SATURN (1999–2010)

To find Saturn, use the map (below) to discover which constellation the planet is in, then use the planisphere to discover the exact location of the constellation for your latitude and the time. Saturn appears like a star to the naked eye. Because it is so far away, the planet moves very slowly across the sky from west to east (right to left on the map) and regularly goes into retrograde motion.

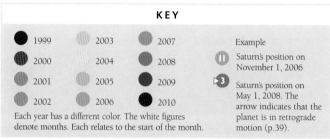

### KEY

| | | |
|---|---|---|
| ● 1999 | 2003 | 2007 |
| 2000 | 2004 | 2008 |
| 2001 | 2005 | 2009 |
| 2002 | 2006 | 2010 |

Each year has a different color. The white figures denote months. Each relates to the start of the month.

Example

Saturn's position on November 1, 2006

Saturn's position on May 1, 2008. The arrow indicates that the planet is in retrograde motion (p.39).

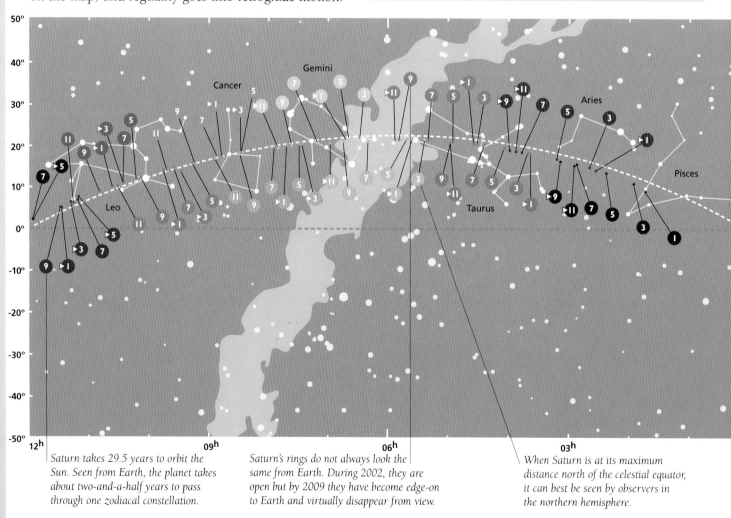

*Saturn takes 29.5 years to orbit the Sun. Seen from Earth, the planet takes about two-and-a-half years to pass through one zodiacal constellation.*

*Saturn's rings do not always look the same from Earth. During 2002, they are open but by 2009 they have become edge-on to Earth and virtually disappear from view.*

*When Saturn is at its maximum distance north of the celestial equator, it can best be seen by observers in the northern hemisphere.*

MERCURY VENUS MARS JUPITER **SATURN** URANUS NEPTUNE PLUTO MOON SUN

*The rings are made of billions of ice-coated pieces of rock.*

*The upper atmosphere has violent storms.*

*The gas clouds form into belts and zones.*

*The equatorial zone is the brightest part of the planet.*

**Scale**
*Saturn is about nine-and-a-half times the diameter of Earth.*

## SATURN'S SURFACE

Saturn, like Jupiter, is a fast-spinning gas giant with a bulging equatorial region and flattened poles. The upper atmosphere, mainly made up of hydrogen, forms into dark belts and bright zones similar to Jupiter's but not so pronounced or colorful when seen from Earth. The broad, bright ring system encircling Saturn can be seen from Earth.

*The tilt of the rings as seen from Earth changes during the planet's orbit of the Sun.*

*The gap in the rings looks empty but spacecraft have shown that there are smaller rings within it.*

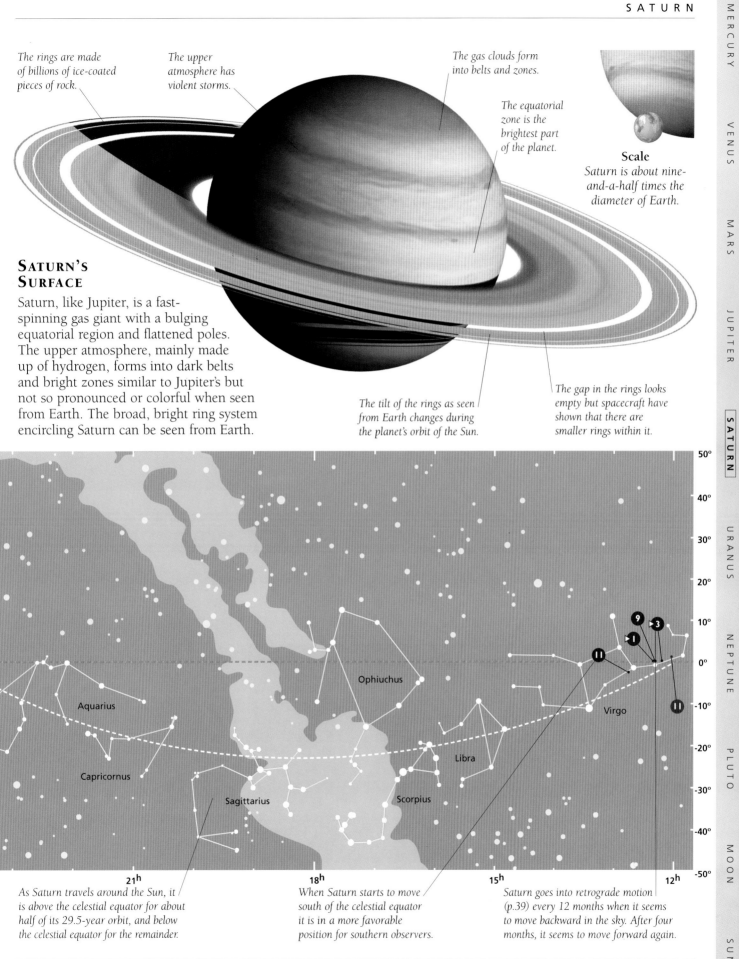

Aquarius

Ophiuchus

Capricornus

Virgo

Libra

Sagittarius

Scorpius

50°
40°
30°
20°
10°
0°
-10°
-20°
-30°
-40°
-50°

21ʰ   18ʰ   15ʰ   12ʰ

*As Saturn travels around the Sun, it is above the celestial equator for about half of its 29.5-year orbit, and below the celestial equator for the remainder.*

*When Saturn starts to move south of the celestial equator it is in a more favorable position for southern observers.*

*Saturn goes into retrograde motion (p.39) every 12 months when it seems to move backward in the sky. After four months, it seems to move forward again.*

Left margin (top to bottom): SUN · MOON · PLUTO · NEPTUNE · URANUS · SATURN · JUPITER · MARS · VENUS · MERCURY

# OBSERVING SATURN

Saturn is a huge planet, but it appears small in the sky because it is so remote. It is, however, bright enough to be spotted with the naked eye, although a telescope is needed to see any surface details. Unlike Jupiter, Saturn has no rapidly changing global features, but observers can detect long-term changes in the visibility and brightness of its belts and zones. A severe weather disturbance can occasionally be seen in its upper atmosphere. Our view of its rings depends upon Earth's position in relation to Saturn's orbit around the Sun.

## LOOKING AT SATURN

**⊙ Naked-eye view**
*Saturn looks like a star. It ranges from mag.–0.3 at its brightest to a fainter 0.8. This is the effect of the rings: when they are face-on to Earth, more light is reflected.*

**◑ Binocular view**
*Through binoculars, Saturn's disk-like shape can be clearly seen. Good binoculars will show the rings, when they are full-on to Earth, as a bump at either side of the planet.*

**✦ Telescope view**
*Seen through a small telescope, Saturn appears as a definite disk and the ring system can be discerned. To see any details, however, a large telescope is needed.*

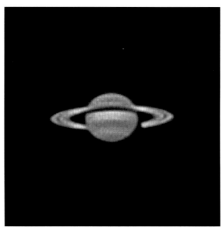

**▣ CCD image view**
*With a CCD camera, the planet takes on a banded appearance and the nature and orientation of the rings as a system separate from the planet now becomes apparent.*

## OUR CHANGING VIEW

As Saturn and Earth move around the Sun, the view we have of the planet and its rings changes. The ring system can be edge-on to our line of sight, or we can view it from below and above. The complete cycle of views takes 29.5 years.

**1999** *Saturn's rings are beginning to open out as the south pole tilts toward the Sun.*

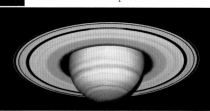

**2002** *The rings are now open. The gap, the Cassini Division, is visible.*

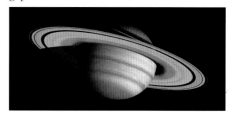

**2006** *Saturn's south pole is tilting away from the Sun. The rings appear to close.*

### BEST TIMES TO OBSERVE

The best time to see Saturn is when the planet is at opposition (p.39). This happens annually about two weeks later each year. The most favorable oppositions will be in 2001 and 2006.

### DATES OF OPPOSITION

| | |
|---|---|
| Nov 6, 1999 | Jan 27, 2006 |
| Nov 19, 2000 | Feb 10, 2007 |
| Dec 3, 2001 | Feb 24, 2008 |
| Dec 17, 2002 | Mar 8, 2009 |
| Dec 31, 2003 | Mar 22, 2010 |
| Jan 13, 2005 | |

**2025** *Once again Saturn's rings are visible edge-on. This event happens only about every 14.75 years.*

**Saturn in the night sky**
*To the naked eye, Saturn (center top) looks indistinguishable from a bright star. To make certain it is Saturn that you are observing, familiarize yourself with the stars in the constellation you know the planet is in.*

Saturn's orbit

Earth

Sun

Line of sight

Earth's orbit

Saturn

**2021** *The gap between Saturn and the start of the rings begins to disappear from view.*

**2017** *The rings have opened out. The north pole tips toward the Sun.*

**2013** *The rings are opening out once again as the north pole tilts toward the Sun.*

**Titan**
*Saturn's moon, Titan, has a nitrogen atmosphere. It is the second largest moon in the entire Solar System.*

**2009** *The planet's ring system is now edge-on to Earth and has almost disappeared from view.*

## SATURN'S MOONS

Saturn has at least 18 moons. Even with a large telescope observers can see fewer than 10. They appear as dots of light circling the planet. The largest, Titan, is the easiest to see.

| Titan | Rhea | Dione | Tethys |

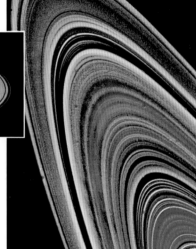

**Enhanced color image of the rings**

## SATURN'S RINGS

Spaceprobes have shown that Saturn is surrounded by a complex system of rings and ringlets, with tiny moons orbiting in the outer rings. The three main rings are visible through a telescope from Earth, as is the gap that separates two of them. This gap is called the Cassini Division, after Giovanni Cassini, who discovered the gap in 1675. It is about 2,980 miles (4,800 km) wide. The three rings may look small from Earth but they are vast. Their diameter would stretch two thirds of the distance from Earth to the Moon. The thickness of the rings is about a mile.

# URANUS, NEPTUNE, AND PLUTO

The outermost planets, Uranus, Neptune, and Pluto are remote, faint, and hard to find. To be sure you are observing them, follow their paths against the background stars. Uranus, closest of the three, can be spotted with the naked eye; fainter and more distant, Neptune can be seen with binoculars or a telescope and Pluto is so tiny, a large telescope is required.

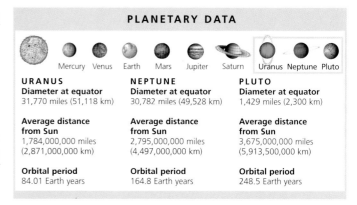

### PLANETARY DATA

Mercury  Venus  Earth  Mars  Jupiter  Saturn  Uranus  Neptune  Pluto

| URANUS | NEPTUNE | PLUTO |
|---|---|---|
| **Diameter at equator** 31,770 miles (51,118 km) | **Diameter at equator** 30,782 miles (49,528 km) | **Diameter at equator** 1,429 miles (2,300 km) |
| **Average distance from Sun** 1,784,000,000 miles (2,871,000,000 km) | **Average distance from Sun** 2,795,000,000 miles (4,497,000,000 km) | **Average distance from Sun** 3,675,000,000 miles (5,913,500,000 km) |
| **Orbital period** 84.01 Earth years | **Orbital period** 164.8 Earth years | **Orbital period** 248.5 Earth years |

## LOCATING URANUS, NEPTUNE, AND PLUTO 1999–2010

Uranus and Neptune stay close to the ecliptic, but Pluto strays. Use the maps below to discover which constellation each planet is in. Use the planisphere to locate the constellation for your latitude and time. At mag.5.5 Uranus is within naked-eye visibility. Neptune at mag.7.8, and Pluto at mag.14.5, are fainter than the stars on these maps.

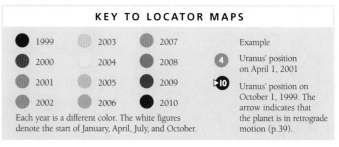

### KEY TO LOCATOR MAPS

| | | | |
|---|---|---|---|
| 1999 | 2003 | 2007 | Example |
| 2000 | 2004 | 2008 | Uranus' position on April 1, 2001 |
| 2001 | 2005 | 2009 | Uranus' position on October 1, 1999. The arrow indicates that the planet is in retrograde motion (p.39). |
| 2002 | 2006 | 2010 | |

Each year is a different color. The white figures denote the start of January, April, July, and October.

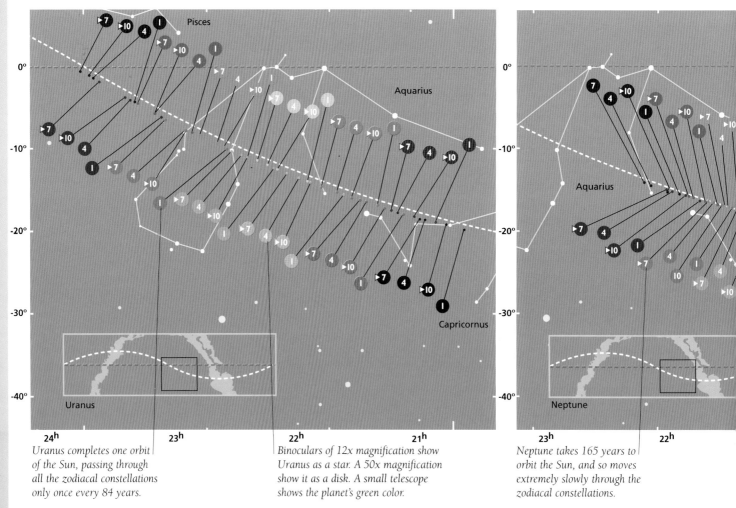

Pisces

Aquarius

Capricornus

Uranus

Neptune

24ʰ  23ʰ  22ʰ  21ʰ  23ʰ  22ʰ

*Uranus completes one orbit of the Sun, passing through all the zodiacal constellations only once every 84 years.*

*Binoculars of 12x magnification show Uranus as a star. A 50x magnification show it as a disk. A small telescope shows the planet's green color.*

*Neptune takes 165 years to orbit the Sun, and so moves extremely slowly through the zodiacal constellations.*

SUN  MOON  PLUTO  NEPTUNE  URANUS  SATURN  JUPITER  MARS  VENUS  MERCURY

## URANUS

Uranus is a large gas planet. Its color comes from clouds of hydrogen, helium, and methane. Uranus is tipped on its side, and has a ring system with a family of moons. The five biggest moons are visible with a large telescope.

## PLUTO

Pluto is so distant and so small that its surface has never been observed from Earth, and a spaceprobe has never visited it. It is believed to be a frozen world of rock and ice. Pluto has one moon, called Charon.

## NEPTUNE

Also a gas giant, Neptune was seen close up by Voyager 2 in 1989. It is similar to Uranus but bluer and smaller. Neptune also has many moons, including Triton.

*Bright wispy clouds of methane ice appear on Neptune's surface.*

*Spaceprobe images showed Uranus as having an almost featureless globe.*

*Neptune has at least four faint rings*

*The Great Dark Spot is a huge storm in the atmosphere.*

**Scale**
*Uranus' diameter is about four times that of Earth.*

**Scale**
*Neptune's diameter is almost four times that of Earth.*

**Scale**
*Pluto's diameter is less than one fifth that of Earth.*

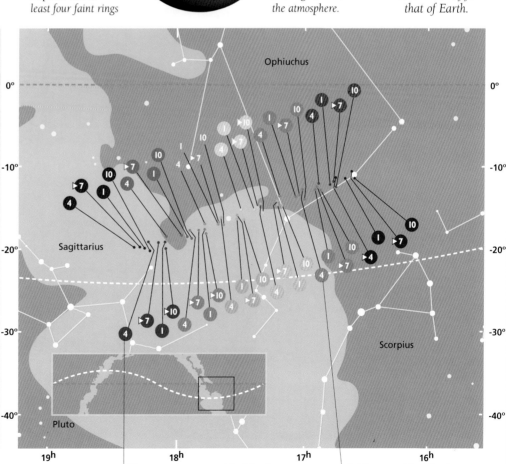

*Binoculars or a small telescope will show Neptune as a starry point of light. A 6-in (150-mm) telescope will show the planet as a blue-green disk.*

*Pluto completes one orbit of the Sun every 248 years. As the planet was first discovered in 1930, only a small part of its path has ever been observed.*

*A telescope with at least an 8-in (200-mm) aperture is needed to see Pluto and even then it appears just like a star.*

# THE MOON

The Moon is Earth's natural satellite, and appears by far the largest object in the night sky. It is a cratered ball of rock about a quarter of the size of Earth and accompanies Earth on its orbit around the Sun. The Moon has no light of its own, but shines through reflected light from the Sun. From day to day, the Moon's disk seems to assume a different shape. These shapes are known as phases and a complete cycle, from New Moon to New Moon, takes 29.5 days.

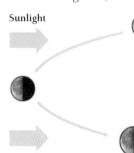

**8 Crescent**
*A thin sliver of sunlit Moon is visible. The Moon is said to be waning, or shrinking in size.*

Sunlight

**1 New Moon**
*The face of the Moon seen from Earth is unlit, and cannot be seen clearly from Earth.*

## LOOKING AT THE MOON

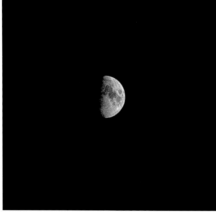

**2 Crescent**
*A thin crescent of one side of the Moon is visible as it is lit by the Sun. The Moon is said to be waxing, or growing.*

**⊙ Naked-eye view**
*Except at New Moon, the Moon is easily spotted in the sky, irrespective of its phase. Dark and light surface features are apparent and it moves quite rapidly across the sky.*

**⊕ Binocular view**
*Through binoculars, the Moon can still be seen as a whole but it now looks closer. More of the Moon's surface features and the uneven nature of the terminator is now visible.*

## THE MOON'S PHASES

As the Moon circles Earth we see differing amounts of its sunlit side. During these phases, the side of the Moon visible from Earth first grows or waxes, and then shrinks or wanes.

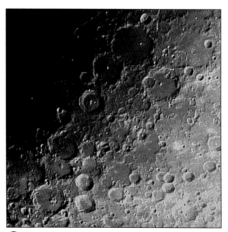

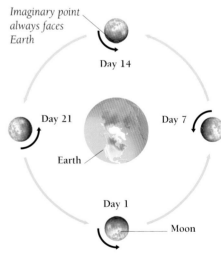

*Imaginary point always faces Earth*

Day 14

Day 21

Day 7

Earth

Day 1

Moon

**✪ Telescope view**
*Through a telescope, details of the Moon's surface are brought into view. Individual features such as craters and seas (maria) are now distinct and easy to observe.*

**⊜ CCD image view**
*The view is now so sharp and clear that the shadows cast by high ground become visible. Fine surface features, such as the central peaks in craters, can also be distinguished.*

## NEAR SIDE OF THE MOON

From Earth we only ever see one side (the near side) of the Moon. This is because the Moon takes exactly the same amount of time to rotate on its own axis as it takes to orbit Earth.

**7 Last Quarter**
*Only half the Moon is illuminated. This phase is called Last Quarter because only a quarter of the phase cycle remains.*

**6 Gibbous**
*The Moon has completed more than half of its phase cycle and is said to be waning gibbous.*

**Scale**
*The Moon's diameter is 2,158 miles (3,476 km), about a quarter of the diameter of Earth.*

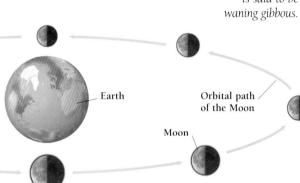

Earth

Orbital path of the Moon

Moon

**5 Full Moon**
*The Moon, now on the opposite side of Earth to the Sun, is fully lit. From Earth we see a complete disk.*

**THE MOON IN THE DAYTIME**

For part of the month, the Moon appears in the daytime sky. At this time it does not shine so brightly, but surface features can be seen.

**3 First Quarter**
*Half the Moon is lit up. The term First Quarter is used because the Moon has completed a quarter of its orbit.*

**4 Gibbous**
*About three quarters of the sunlit side of the Moon can be seen. This phase is called waxing gibbous.*

**THE TERMINATOR**

The boundary between the sunlit part and the dark part of the Moon is called the terminator. Features can be seen to best effect along the line of the terminator as it moves across the Moon's face with its changing phases.

*Craters and mountains are thrown into relief along the terminator.*

**THE MOON IN THE SKY**

Traditionally, Moon maps have north at the top. Observers in the southern hemisphere, however, will see the south pole at the top when they look at the Moon in the sky.

**Full Moon**
*The Moon is bright but its features appear indistinct in the strong, direct sunlight, and Full Moon is not a good time for observing.*

### LUNAR DATA

| | |
|---|---|
| **Diameter at equator** 2,160 miles (3,476 km) | **Orbital period** 27.3 Earth days |
| **Average distance from Earth** 238,774 miles (384,500 km) | **Gravity (Earth=1)** 0.16 |
| **Surface temperature** –311° to 221°F (–155° to 105°C) | **Mass (Earth=1)** 0.012 |
| **Rotation period** 27.3 Earth days | **Interval between New Moons** 29.5 Earth days |

# OBSERVING THE MOON 1

The Moon is a lifeless, rocky world without air or water. The lunar landscape has remained virtually unchanged for millions of years. During its early formation, the Moon was bombarded by space rocks that formed craters. The impact of the rocks created mountains. In time, lava seeped from inside the Moon and filled some of the craters; this produced dark, flat areas that looked very much like seas, so they were called maria, Latin for seas (sing. mare).

**Mare Imbrium**
*About 3.8 billion years ago, this dark flat plain was a gigantic crater about 700 miles (1,000 km) wide, but lava flooded it. More recent craters such as Archimedes formed within it. Mare Imbrium can be seen with binoculars or a telescope.*

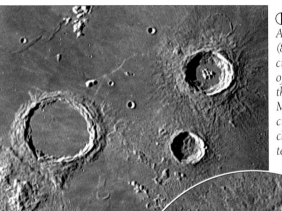

**Archimedes**
*At about 50 miles (80 km) across this crater is the largest of three craters on the eastern edge of Mare Imbrium. The crater walls are all clearly visible in this telescope view.*

**Copernicus**
*This ray crater can be seen with the naked eye. Around the 60-mile (97-km) outer ring are the rays formed by material ejected from the crater. Binoculars will bring the crater into view. Its high, terraced walls and central mountains can be seen clearly through a telescope.*

*Tycho is a ray crater that makes an impressive sight through binoculars.*

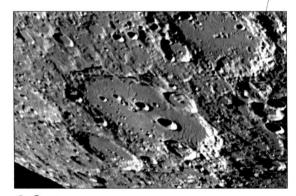

**Clavius**
*One of the largest craters on the Moon, Clavius, measures about 142 miles (230 km) across. Within its walls there are several smaller craters.*

Montes Jura
Sinus Iridum
Mare Imbrium
Montes Apenninu
Montes Carpatus
Oceanus Procellarum
Mare Cognitum
Mare Humorum
Mare Nubium

19
14
20
21
18
22
24
23
1
15
16
26
25
17

North

The first Moon walk took place in the Mare Tranquillitatis in 1969.

Mare Frigoris

Montes Alpes

Montes Caucasus

Mare Serenitatis

Mare Vaporum

Mare Crisium

Mare Tranquillitatis

Mare Fecunditatis

Mare Nectaris

South

◑ ⚛ **Ptolemaeus**
A row of three craters to the left of the picture shows Ptolemaeus at the top with Alphonsus and Arzachel below it. A large telescope shows that Ptolemaeus is covered in smaller craters.

**Theophilus** ⚛ ◑
This 62-mile (101-km) crater is found at the north-eastern corner of Mare Nectaris. Its terraced walls rise more than 16,400 feet (5,000 meters) above the crater floor. A mountain range is at the center of the crater.

Petavius is a large crater with walls that rise to over 11,500 feet (3,500 meters)

Stöfler has a dark-gray floor that makes it easy to spot. Part of its wall has been breached by a smaller crater.

### BEST TIMES TO OBSERVE

| | | |
|---|---|---|
| ◐ Waxing crescent | | ◑ Last quarter |
| ◑ First quarter | | ◐ Waning crescent |

### WHAT TO OBSERVE

Uplands, lowlands, and maria are named on the central map and all are visible with binoculars. Below is a selection of craters that you can see with binoculars. The symbols suggest the phase when the crater is close to the terminator.

| 1 ◐ Endymion | 14 ◑ Plato |
|---|---|
| 2 ◐ Posidonius | 15 ◑ Alphonsus |
| 3 ◐ Langrenus | 16 ◑ Tycho |
| 4 ◐ Cleomedes | 17 ◑ Maginus |
| 5 ◐ Petavius | 18 ◑ Eratosthenes |
| 6 ◐ Janssen | 19 ◐ Pythagoras |
| 7 ◐ Atlas | 20 ◐ Aristarchus |
| 8 ◑ Aristoteles | 21 ◐ Kepler |
| 9 ◑ Albategnius | 22 ◐ Grimaldi |
| 10 ◑ Aliacensis | 23 ◐ Bullialdus |
| 11 ◑ Ptolemaeus | 24 ◐ Gassendi |
| 12 ◑ Aristillus | 25 ◐ Schiller |
| 13 ◑ Stöfler | 26 ◐ Longomontanus |

## SKETCHING THE MOON

Sketching the Moon's surface will help to train your eyes to observe astronomical detail, and in the process you will become very familiar with the Moon. Observing with the naked-eye or binoculars, you can draw the light and dark features of the lunar surface. You can also record the Moon's phases and the prominent markings along the terminator at each phase. Observers with a telescope can sketch individual craters and maria on the lunar surface.

Use a notebook and soft pencil.

# OBSERVING THE MOON 2

The Moon is always fascinating to observe. Although we can only ever see one side of it, this side is constantly changing, because sunlight strikes it in different ways as it goes through its phases (pp. 62–3). Occasionally, light from the Sun is prevented from reaching the Moon and a lunar eclipse occurs. During an eclipse, the Moon darkens, but instead of disappearing from view, it assumes a colorful, reddish glow that can often be spectacular. Sometimes, an interesting effect called Earthshine can be seen on the Moon. This happens when sunlight reaching Earth reflects onto the surface of the Moon.

## LUNAR ECLIPSE

A lunar eclipse occurs when the Moon passes into the shadow cast by the night-time side of Earth. Sometimes the Moon is totally covered by the shadow (a total eclipse) and sometimes only partly (a partial eclipse). The Moon is eclipsed up to three times a year. No special equipment is needed to observe a lunar eclipse.

*The color and brightness of the Moon's surface depends on Earth's atmosphere*

*Earth's shadow can be seen on the Moon.*

**Lunar eclipse: stage 1**
*The Full Moon is moving across Earth's sky. Its lower edge has entered Earth's shadow.*

*Only half of the Moon's face is lit by the Sun.*

**Lunar eclipse: stage 2**
*About half of the Moon's disk is in Earth's shadow. The Moon is reddening.*

*Only a thin sunlit crescent remains.*

**Lunar eclipse: stage 3**
*The Moon is almost totally eclipsed; soon Earth's shadow will cover the Moon.*

## HOW A LUNAR ECLIPSE HAPPENS

When the Sun, Earth, and the Moon are aligned, the Earth's shadow prevents sunlight from reaching the Moon. When the Moon is in the darkest part of the shadow, it receives no direct light and a total eclipse occurs. When only part of the lunar disk is in the darkest part of the shadow, a partial eclipse occurs.

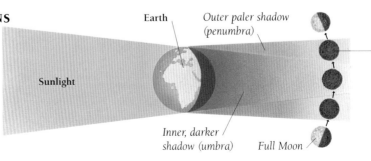

Earth

*Outer paler shadow (penumbra)*

Sunlight

*Only a slight darkening of the Moon may occur in the light outer shadow*

*Inner, darker shadow (umbra)*

*Full Moon*

MERCURY VENUS MARS JUPITER SATURN URANUS NEPTUNE PLUTO SUN MOON

*Light bending round Earth, through its atmosphere, can lighten the face of the Moon*

## THE FAR SIDE OF THE MOON

Thanks to pictures taken by spacecraft, we are familiar with the far side of the Moon that we can never see from Earth. There are numerous craters, but no large maria. We do actually see a little of the far side from Earth, as 59 percent of the Moon is visible as the Moon completes one orbit of Earth.

*Tsiolkovsky is a 112 mile (180 km) crater with a dark floor.*

*Mare Moscoviense is one of the few maria on the far side of the Moon*

## EARTHSHINE

Both Earth and the Moon shine because light from the Sun falls on them. An effect known as Earthshine occurs when light can be seen reflected on to the unlit part of the Moon from Earth. This happens close to the New Moon period.

**Total Eclipse**
*During an eclipse, the Moon can become red in color. This total lunar eclipse occurred on December 9, 1992.*

### LUNAR ECLIPSES

A total lunar eclipse has been known to last for as long as one hour 47 minutes. A lunar eclipse always looks the same from wherever it is viewed on Earth and provided that the Moon is above the horizon and the sky is sufficiently cloudless, it should be clearly visible.

### DATES OF LUNAR ECLIPSES

The dates below are for future total and partial eclipses seen throughout the world.

| Total eclipse | Feb 21, 2008 |
|---|---|
| Jan 21, 2000 | Dec 21, 2010 |
| Jul 16, 2000 | **Partial eclipse** |
| Jan 9, 2001 | Jul 28, 1999 |
| May 16, 2003 | Jul 5, 2001 |
| Nov 8–9, 2003 | Oct 17, 2005 |
| May 4, 2004 | Sep 7, 2006 |
| Oct 28, 2004 | Aug 16, 2008 |
| Mar 3–4, 2007 | Dec 31, 2009 |
| Aug 28, 2007 | Jun 26, 2010 |

## LUNAR OCCULTATIONS

The Moon moves rapidly across Earth's sky. As it does so it passes in front of stars and planets. Even though these celestial bodies are at a much greater distance than the Moon, for a short time, as the Moon crosses over them, they lie in the same line of sight and are occulted by the Moon.

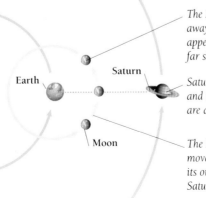

*The Moon moves away Saturn appears on the far side of it.*

Saturn

Earth

*Saturn, Earth, and the Moon are aligned.*

Moon

*The Moon moves along its orbit toward Saturn.*

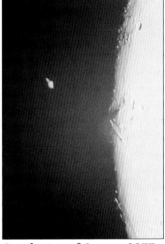

**Occultation of Saturn, 1977**
*An occulted object, in this case Saturn (pp.56–9), can be observed as it disappears from view behind the edge of the Moon and then appears again on the other side.*

MERCURY VENUS MARS JUPITER SATURN URANUS NEPTUNE PLUTO **MOON** SUN

# THE SUN

The Sun is the closest star to Earth, and the only one that can be observed in any detail. It is a giant incandescent ball consisting mainly of hydrogen. At the Sun's core, hydrogen is converted to helium at colossal temperatures and this produces the energy that we experience on Earth as light and heat. The Sun's brilliant light makes it a dangerous object to observe, but with care some of its surface details can be seen.

## SOLAR DATA

| | |
|---|---|
| **Average diameter** 864,000 miles (1,392,000 km) | **Rotation period at poles** 35 Earth days |
| **Average surface temperature** 9,900°F (5,500°C) | **Distance from Earth** 93,000,000 miles (149,680,000 km ) |
| **Rotation period at equator** 25 Earth days | **Mass (Earth=1)** 330,000 |

## OBSERVING THE SUN SAFELY

The only way to observe the Sun is with protection, or by projection. Filters for telescopes are available (p.27) but the projection method is better because it does not involve looking directly at the Sun and eliminates any risk.

*The image of the Sun is directed onto the card.*

**Using binoculars**
*The eyepiece end of binoculars can be used to direct the image of the Sun on to a piece of card which acts as a screen.*

**Step 1**
*Position a piece of card about 18 in (50 cm) from the eyepiece. Cap the finder, for safety.*

**Step 2**
*Move the eyepiece lens gently in toward tube of the telescope until the image is sharp.*

**Using a telescope**
*If you have a telescope, it can be used to direct an image of the Sun safely through the eyepiece onto a viewing screen. To accomplish this, aim the objective lens of the telescope directly at the Sun.*

## THE SUN'S SURFACE

The bright, outer layer of the Sun that we see from Earth is called the photosphere. It looks yellow because of the temperature of the surface, which is about 9,900°F (5,500°C). Although the Sun spins fast for its size, it does not bulge at the equator and retains its perfectly spherical shape.

Earth

**Scale**
*The Earth is tiny compared with the Sun. Some 109 Earths would fit across its face and 1.3 million would be needed to fit inside it.*

> **WARNING**
> Never look at the Sun directly with the naked eye, or with any instrument. The light will burn your retina, causing permanent blindness.

*Above the photosphere is a layer called the chromosphere.*

## SUNLIGHT AND EARTH'S ATMOSPHERE

As the Sun's light passes through Earth's atmosphere, it can be broken up into different wavelengths. At sunset, the blue wavelengths are scattered and red predominates.

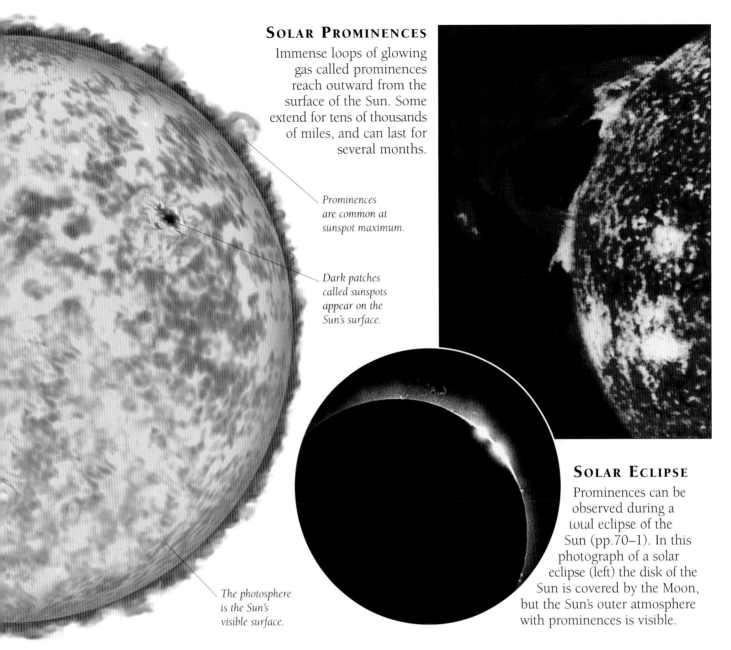

## SOLAR PROMINENCES

Immense loops of glowing gas called prominences reach outward from the surface of the Sun. Some extend for tens of thousands of miles, and can last for several months.

*Prominences are common at sunspot maximum.*

*Dark patches called sunspots appear on the Sun's surface.*

*The photosphere is the Sun's visible surface.*

## SOLAR ECLIPSE

Prominences can be observed during a total eclipse of the Sun (pp.70–1). In this photograph of a solar eclipse (left) the disk of the Sun is covered by the Moon, but the Sun's outer atmosphere with prominences is visible.

## SUNSPOTS

Sunspots, the dark patches that appear on the Sun's surface, are relatively cool regions of the photosphere. They are large—more than 30,000 miles (50,000 km) across—and usually appear in pairs or groups. As the Sun rotates, the sunspot groups appear to travel across the Sun's surface. They move from one edge of the Sun to the other in about two weeks. Spots appear between 40ºN and 40ºS of the equator.

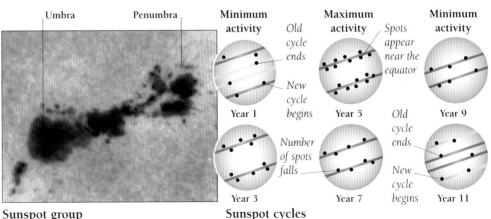

Umbra    Penumbra

**Sunspot group**
*The umbra is about 2,732°F (1,500°C) cooler than the solar surface. Around this umbra is the lighter, hotter penumbra.*

**Minimum activity**
*Old cycle ends*
Year 1
*New cycle begins*

**Maximum activity**
*Spots appear near the equator*
Year 5

**Minimum activity**
Year 9

Year 3
*Number of spots falls*

Year 7

*Old cycle ends*
*New cycle begins*
Year 11

**Sunspot cycles**
*Sunspot activity follows an 11-year cycle. At maximum, about 120 spots per month are visible on the solar surface. At minimum, this number drops to about six.*

SUN MOON PLUTO NEPTUNE URANUS SATURN JUPITER MARS VENUS MERCURY

# SOLAR ECLIPSE

Although the Sun and Moon are vastly different in size and distance from Earth, by an odd coincidence they appear the same size in the sky and, when the three bodies are in alignment, the Moon will cover the entire disk of the Sun and completely block it from view. This is called a solar eclipse. A solar eclipse occurs only at New Moon (p.62), when the Moon's shadow falls on Earth. Eclipses do not happen every New Moon; in fact they are rare and dramatic events, occurring only once or twice a year, and they are only visible from parts of Earth. Only observers in the inner part of the Moon's shadow will see a total eclipse.

*The Sun's outermost gaseous layer, the corona, can be seen only when the disk is eclipsed.*

*The New Moon covers the disk of the Sun.*

*The Sun's corona is a million times fainter than the solar disk.*

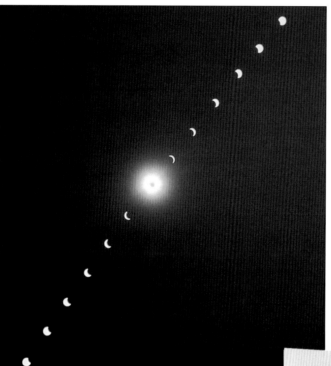

### TIME-LAPSE IMAGES OF AN ECLIPSE
These images were taken at 10-minute intervals, starting at bottom left. It took about 70 minutes for the Moon to cover the Sun. The central image shows totality, or total eclipsing of the Sun, which lasts up to 7 minutes 40 seconds.

### HOW A SOLAR ECLIPSE HAPPENS
When the Moon is directly between the Sun and Earth, it blots out the Sun and casts a shadow on Earth. The inner part of the shadow is called the umbra and it can be about 100 miles (160 km) wide. Observers in the umbra see a total eclipse; those in the penumbra see a partial eclipse.

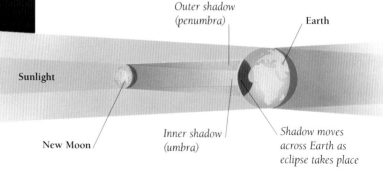

*Outer shadow (penumbra)*

*Earth*

Sunlight

New Moon

*Inner shadow (umbra)*

*Shadow moves across Earth as eclipse takes place*

## OBSERVING SOLAR ECLIPSES

A solar eclipse can only be seen from the parts of Earth covered by the Moon's shadow and only appears total inside the umbra; from within the penumbra the eclipse is partial. As the eclipse takes place, the shadow moves in a path across the surface of Earth. The path of the shadow on Earth is different for each eclipse. The dates below are for forthcoming eclipses that can be seen from around the world.

### DATES OF TOTAL ECLIPSES

| | |
|---|---|
| Aug 11, 1999 | Mar 29, 2006 |
| Jun 21, 2001 | Aug 1, 2008 |
| Dec 4, 2002 | Jul 22, 2009 |
| Nov 23, 2003 | Jul 11, 2010 |
| Apr 8, 2005 | |

## TOTAL SOLAR ECLIPSE

Athough the Moon is 400 times smaller than the Sun, it is 400 times closer to Earth, and so the two bodies appear roughly the same size, about half a degree across, in the sky. Total eclipses occur when the Moon's disk completely covers the Sun and the outer halo of gas, the corona, becomes visible.

## DIAMOND RING EFFECT

Just as totality begins and just as it ends, a spectacular effect occurs that looks similar to a diamond set in a ring. The effect is caused by small portions of the Sun remaining visible between the mountains on the surface of the Moon. It lasts just a few seconds.

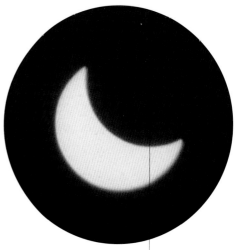

*Observers south of the path see the Moon cover the top of the Sun*

## PARTIAL ECLIPSE

Observers outside the narrow path of totality, but within the lighter outer shadow, the penumbra, see a partial eclipse. Although not as dramatic as a total eclipse, it is still a wonderful sight, as the Sun appears to have had a bite taken out of it, as the Moon crosses over it.

## ANNULAR ECLIPSE

**A ring of light is seen around the Moon**

Outer shadow (penumbra) — Moon — Inner shadow (umbra) — Earth — Sunlight

*Area covered by a total eclipse* — *Area covered by a partial eclipse*

An annular eclipse occurs when the Moon does not completely cover the Sun's disk, and there is still a bright ring, or annulus, of light remaining. This happens when the Moon is at its furthest distance from Earth. The effect is disappointing because the Sun's corona can not be observed.

# AURORAE

Aurorae (sing. aurora) are spectacular displays of colored rays, streamers, and arcs in the sky. The red, pink, green, and blue lights originate in the atmosphere above Earth's polar regions. Anyone located in the most northerly and southerly latitudes of Earth can see them—they are easy to spot and watch with the naked eye. Aurorae can happen at any time and are difficult to predict, but on average, there is a display every month in the northern and southern hemispheres.

### AURORA BOREALIS

This photograph was taken in the early morning of May 1, 1990 in Scotland. The display started in the late evening the day before, and went on through the night until it vanished in the dawn light. The aurora changed over that time, showing single rays, streamers, and arcs. Here, a folding arc has rays extending upward toward the constellation Cassiopeia.

**Observing aurorae**
*The Aurora Borealis (Northern Lights) can be seen from locations north of about latitude 50°N. The Aurora Australis (Southern Lights) can be seen south of about 50°S. This diagram shows the most populated areas where aurorae are likely to be seen.*

- Aurora Borealis
- Aurora Australis

### HOW DOES AN AURORA HAPPEN?

Charged particles originating in the Sun and carried on the solar wind enter Earth's upper atmosphere and are attracted to Earth's poles. The particles spiral along Earth's magnetic field lines toward the north and south polar regions. Gas in the atmosphere interacts with the particles and glows, creating spectacular light displays.

Auroral effect

Earth's atmosphere

Night

Day

*Charged particles from the Sun are drawn to the north pole.*

*Charged particles from the Sun are drawn to the south pole.*

## AURORA AUSTRALIS

The Aurora Australis is never seen by as many people as the Aurora Borealis, because the light displays occur mainly over uninhabited land. This photograph of a colorful green aurora was taken from Cape Evans in Antarctica with a 35-mm camera and a fast film that was set for a long exposure.

## SHAPES OF AURORAE

Arcs and rays can be seen in auroral displays. A folding arc, resembling ribbon folds, is called a band, and rays stretching up like curtains are known as a rayed band. Colors depend on the height of the aurora and the composition of Earth's atmosphere.

**Aurora Borealis, March 25, 1991**
*A crown of light, or auroral corona, is seen when rays appear to stretch upward from the ground, along the entire horizon.*

**Aurora Borealis, February 15, 1990**
*The green-yellow colors are caused by oxygen in the atmosphere at 56–93 miles (90–150 km) above the ground.*

**Aurora Borealis, May 1, 1990**
*Red light is caused by oxygen at more than 93 miles (150 km) and from hydrogen at about 74 miles (120 km) above the ground.*

**Aurora Borealis, December 26, 1989**
*Rays are the most common shape—here they stretch up in columns from the horizon. They can be seen moving from left to right.*

# METEORS

Dust and rock from space are continuously bombarding Earth. More than 200,000 tons enters our atmosphere each year. The dust and rock pieces are called meteoroids, and most burn up in Earth's atmosphere. As a meteoroid burns itself out, it produces a meteor—a short-lived trail of light. These are popularly called shooting stars. At certain times, showers of meteors can be seen raining down.

*A single meteor that is not part of a shower is called a sporadic meteor.*

*The Leonids meteor shower occurs in mid-November every year.*

## METEOR SHOWERS

When Earth passes through the debris left by a comet, a meteor shower occurs. All the meteors in a shower seem to originate from the same point in the sky, known as the radiant. A shower takes its name from the constellation that includes its radiant; for example, the Leonids (above) originate in Leo.

## SINGLE METEOR

The streak of light of a single meteor— either falling on its own or as part of a shower— lasts for less than a second. The average magnitude of a meteor is about 2.5. Single meteors can occur on any night of the year.

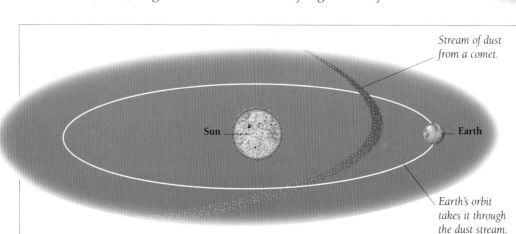

*Stream of dust from a comet.*

Sun

Earth

*Earth's orbit takes it through the dust stream.*

## HOW DOES A METEOR SHOWER HAPPEN?

The dust and rock that causes meteors comes from comets or, sometimes, asteroids. Particles given off by a comet as it travels close to the Sun form a stream of material. When Earth travels through this stream on its orbit of the Sun, the particles enter Earth's atmosphere and a meteor shower is produced.

## METEORITES

Pieces of space rock, too large to burn up as they enter Earth's atmosphere, reach the planet's surface. These are known as meteorites. More than 3,000 of them, each weighing 2 lbs (1 kg) or more, come through the atmosphere and land on Earth each year. Many fall into the sea, which covers much of Earth. On average about six are seen to fall on land and are collected. Meteorites can form large craters on Earth's surface. More than 150 craters have been identified around the world— some formed millions of years ago, others far more recently.

### Composition
*Meteorites are formed from rock, iron, or a mixture of the two. Rocky meteorites are the most common type to have been found.*

*Rare meteorites contain metal and rock.*

### Meteorite crater
*There are craters on every continent on Earth. This one, 0.8 miles (1.3 km) across, is in Arizona, US, and was formed about 25,000 years ago.*

## FIREBALLS

Meteors of significant brightness sometimes occur; they are called fireballs. Fireballs are produced by meteoroids that are larger than those creating normal meteors. Sometimes, a fireball meteoroid is too big to burn out as it enters Earth's atmosphere and falls to Earth as a meteorite (see above).

*Fireball seen from the Channel Islands during the Perseids of August, 1991. The stars of Ursa Minor are visible in the background.*

### Geminid meteors
*In December each year one of the best meteor showers, the Geminids, occurs. The origin of the shower is an asteroid— Phaethon. Geminid meteors, like all meteors, form at around 43–71 miles (70–115 km) above ground and are typically 2–12 miles (7–20 km) long. At Geminid maximum—when the highest rate of meteors is produced—up to 80 an hour can be seen.*

### ANNUAL METEOR SHOWERS

Meteor showers fall at the same time every year. Use your planisphere to see if the appropriate constellation is visible.

| NAME | DATE |
| --- | --- |
| Quadrantids (Boötes) | Jan 1–6 |
| April Lyrids | Apr 19–24 |
| Eta Aquarids | May 1–8 |
| Delta Aquarids | Jul 15–Aug 15 |
| Perseids | Jul 25–Aug 18 |
| Orionids | Oct 16–27 |
| Taurids | Oct 20–Nov 30 |
| Leonids | Nov 15–20 |
| Geminids | Dec 7–15 |

# COMETS

Comets are lumps of snow and dust, or "dirty snowballs," just a few miles in size. They are leftovers from the formation of the Solar System, and there are millions of them at the very fringes of the present Solar System. Every year, however, some comets travel in toward the Sun, coming close enough to Earth for us to see. By the time a comet is visible from Earth, it has undergone many dramatic changes and become a spectacular object with a glowing head and long tail. Some comets return to our skies in a relatively short period of time, less than 200 years, and are called periodic comets, but others may take tens of thousands of years to return, if at all.

## THE COMET'S TAIL

As a comet nears the Sun, the surface of the nucleus heats up, and the snow changes from a solid to a gas, releasing the dust. This material becomes a giant cloud, called the coma. As the comet gets close to the Sun, the material forms into a long tail. When the comet moves away from the Sun, the coma and tail disappear.

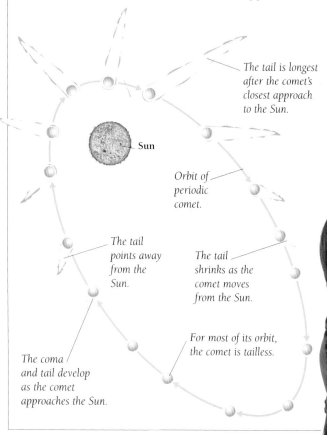

The tail is longest after the comet's closest approach to the Sun.

Sun

Orbit of periodic comet.

The tail points away from the Sun.

The tail shrinks as the comet moves from the Sun.

For most of its orbit, the comet is tailless.

The coma and tail develop as the comet approaches the Sun.

Two tails form, one of gas and one of dust, but the two are not always distinct.

The coma can be many times the diameter of Earth.

The tiny nucleus is hidden from view in the coma's center.

## LOOKING FOR COMETS

Experienced observers use giant binoculars to sweep the night sky for new comets. They search painstakingly for unfamiliar fuzzy stars. If they discover a new comet, it will bear their name. Less powerful binoculars (left) allow a close look at a handful of comets that appear each year.

A comet's tail can be millions of miles long.

## STRUCTURE OF A COMET

A comet consists of a solid nucleus —a snow and dust mixture with a surface of dark dust. As the comet nears the Sun, a bright head of gas and dust forms, called the coma, and from this an extensive tail streams out. Eventually, the comet loses its head and tails.

## PERIODIC COMETS

Some comets return every few years. Below are four to view with the naked eye or binoculars in the years ahead. A more complete list detailing when to see comets for 1999 to 2010 is on pp.136–7.

**Honda-Mrkos-Pajdusakova**
*This comet will be seen by northern and southern observers in April 2001 and June 2006.*

**Borrelly**
*Borrelly will be visible for northern and southern observers from August to November 2001.*

## COMET HALE-BOPP

Comet Hale-Bopp was discovered in 1995 by astronomers Alan Hale and Thomas Bopp. In 1997, it was clearly visible even to the naked eye as it journeyed round the Sun. Astronomers think that Hale-Bopp will not be back for 2,400 years.

**Schwassmann-Wachmann 3**
*This fragmented comet is due in January 2001 and May 2006, when it will come close to Earth.*

---

### TIPS FOR OBSERVERS

**A comet looks** like a fuzzy patch of light to the naked eye; an elongated patch suggests the presence of a tail.
**If you have trouble** seeing the comet, or can see the comet's head but not its tail, use averted vision (p.15).
**If you are not sure** the object is a comet, check its movement against the starry background. You will not see it move but you can plot its nightly progress.
**Observe changes** in brightness and length, and look carefully to see if you can distinguish two tails.

## COMET WEST

Comets enter and leave the inner Solar System at any angle. Comet West, discovered in 1975 by Richard West, arrived perpendicular to the planetary plane. With its broad yellowish-white dust tail and bluish gas tail, it was seen to best effect in 1976 (above). Astronomers think it will not return for about 300,000 years, if at all.

**Kopff**
*Kopff will be visible to both northern and southern hemisphere observers in 2009.*

# ASTEROIDS

The Solar System contains millions of space rocks, called asteroids. Only one, Ceres, is bigger than 560 miles (900 km) across; the smallest are mere specks of dust. Most asteroids exist in a broad band known as the Asteroid, or Main, Belt that lies between the orbits of Mars and Jupiter. Like planets, they rotate as they orbit the Sun. Some stray from the Asteroid belt and cross the paths of the inner plants. All asteroids shine by reflected sunlight but only Vesta, the third largest, is bright enough to be seen with the naked eye.

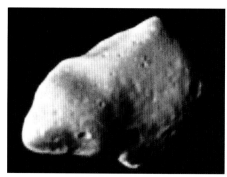

**Close-up photograph of Gaspra**
*Gaspra was the first asteroid to be seen at close quarters. This photograph was taken in October 1991 by the Galileo spacecraft. The asteroid is just 8 miles (12 km) across and is on the very edge of the Asteroid Belt.*

## THE ASTEROID OR MAIN BELT

More than 90 per cent of asteroids exist in the Asteroid Belt, and some 7,000 of these have been identified. They can, however, also be found in other parts of the Solar System. A group known as the Trojans follows the orbit of Jupiter. Asteroids take between three and six years to orbit the Sun.

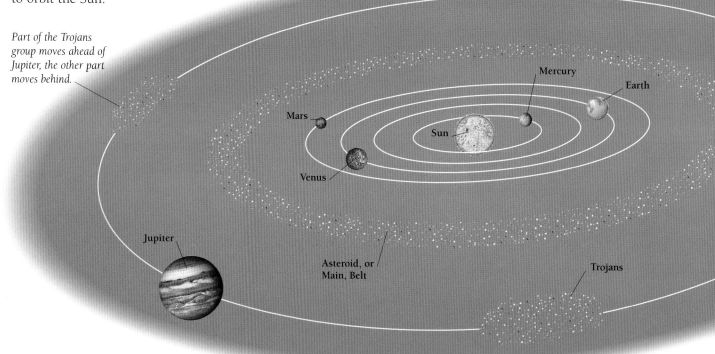

*Part of the Trojans group moves ahead of Jupiter, the other part moves behind.*

Mercury

Earth

Mars

Sun

Venus

Jupiter

Asteroid, or Main, Belt

Trojans

## SHAPES AND SIZES

Asteroids vary in shape and size. Only about ten are larger than 155 miles (250 km). Most are irregularly shaped and only those over 186 miles (299 km) are round. Asteroids are made of rock, metal, or both.

**Psyche**
*An irregularly shaped Main-Belt asteroid, Psyche is about 155 miles (250 km) long. It is believed to be made of stony-iron and has a rotation period of roughly four hours.*

**Vesta**
*A spherical Main-Belt asteroid, Vesta is about 350 miles (560 km) across and is the third largest asteroid known. Its entire surface is pitted with impact craters.*

**Ceres**
*At 580 miles (933 km) across, Ceres is the largest asteroid and was the first to be discovered. It is covered with dark matter that reflects little sunlight.*

**Toutatis: A fast-moving asteroid**
*Discovered in January 1989, Toutatis is a member of a group of asteroids that cross Earth's orbit. Toutatis's speedy progression across the night sky produced the light trail on this long-exposure photograph.*

## VESTA

Photographs are used to confirm the sighting of an asteroid. Vesta is the only one that is bright enough to be seen with the naked eye. At its brightest, Vesta reaches mag.5.5. Its surface reflects a higher proportion of sunlight than any other asteroid. Even so, it only ever appears as a speck of light. A comparison of the same starfield taken at different times will show its movement against the star background.

**Vesta 1**
*Vesta is indistinguishable from the background stars.*

**Vesta 2**
*Two nights later, Vesta has moved to the right.*

## ARTIFICIAL OBJECTS

Some bright streaks of light recorded on film are made by artificial objects orbiting Earth. The Mir space station and the Space Shuttle, for instance, have shiny metal surfaces that reflect sunlight and although these objects are small and hundreds of miles above Earth, they can be detected and followed across the night sky.

**Satellite trail**
*At any one time, there are several hundred artificial satellites in the sky. In this photograph, a satellite has left a trail of light (the dashed lines) as it moved rapidly across the sky.*

**Mir space station**
*Man-made objects, such as the Mir space station (left), can be seen with the naked eye or binoculars as they move rapidly across the sky. They take only minutes to cross from one horizon to the other. The best times to look for them is before dawn and after sunset.*

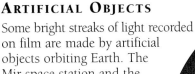

### TIPS FOR OBSERVERS

**Viewed from Earth,** asteroids look like points of light.
**There are more** than 60 asteroids of mag.10 or brighter. This is bright enough for binoculars or a telescope.
**You will need** detailed star maps of faint stars and coordinates of the asteroid you wish to see.
**Compare the star map** with the stars in the sky. Is there an extra starry object? This could be an asteroid.
**Observe the object** over several nights to determine its movement against the starry background.

STARS EXIST IN GREATER NUMBERS IN THE UNIVERSE THAN ANYTHING ELSE. THIS SECTION DESCRIBES THE NATURE OF STARS AND INTRODUCES THE STAR PATTERNS. THE ENTIRE SKY IS SHOWN IN A SERIES OF MAPS, WITH INDIVIDUAL MAPS OF THE CONSTELLATIONS.

# OBSERVING
# *the* STARS

# PATTERNS IN THE SKY

Wherever you are on Earth's surface, you can look up and see stars in the night sky. At first glance, all stars appear much the same and distinguishing one star from another initially seems impossible. However, when you persevere, some stars seem brighter than others and, by joining these together, certain shapes emerge. These patterns, the constellations, have been used since antiquity and still provide the best means of learning your way around the sky.

Equuleus

Aquarius

## YOUR VIEW

All stars are so distant from Earth that they appear as tiny pinpoints of light, whether you use naked eye, binoculars, or a telescope to view them.

### 👁 Naked eye
*Hundreds to thousands of stars can be seen with the naked eye on a clear, dark night. Bright stars, like Polaris, are easiest to view.*

### 🔭 Binoculars
*Stars still look like pinpoints of light through binoculars, only now more can be seen. The fainter stars close to Polaris become visible.*

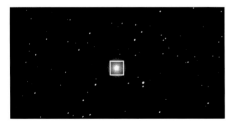

### ✦ Telescope
*A telescope view will show more stars in the same part of sky as Polaris, and Polaris itself can be seen as a double star (p.86).*

## THE CONSTELLATIONS

A constellation consists of a star pattern and the sky immediately around it. Start to find your way about the sky by learning a few, then more, of these star patterns.

Capricornus

*Some star patterns are easier to understand than others. This one is Capricornus, a sea goat—a goat with a fish's tail.*

SCP

### Octans
*This constellation contains the south celestial pole (SCP) Its shape represents a navigator's octant.*

*The constellations fit together like a giant jigsaw puzzle.*

## JIGSAW OF CONSTELLATIONS

The sky surrounding Earth is divided into 88 internationally agreed constellations. These fit together like a giant jigsaw puzzle to form a sphere around Earth. They are of various sizes and consist of different numbers of stars. The first patterns were formed about 4,000 years ago.

Aquila  The bright star, Altair, draws the eye to Aquila which is easily recognizable as an eagle flying across the sky.

Some star patterns represent mythical creatures, including Sagittarius, the centaur (half man, half beast), aiming his arrow at Scorpius.

Scutum

Sagittarius

## OTHER STAR-LIKE OBJECTS

On closer inspection, some pinpoints of light that are seen in the night sky may not turn out to be stars at all. The planets can be easily mistaken for bright stars, and some fuzzy points of light are not, in fact, individual stars but star clusters and galaxies—collections of large numbers of stars.

**Star clusters**
*Star clusters are vast collections of hundreds or even several thousands of stars that form distinct groups.*

**Galaxies**
*Galaxies contain millions of stars. Seen only as a blur, the Andromeda Galaxy is spiral shaped.*

**Planets**
*A planet can appear like a star to the naked eye. Look again and you will see that the planet is disk-like and not a pinpoint of light.*

## STAR DISTANCE WITHIN A CONSTELLATION

The stars in the sky are so far away that they all appear to be the same distance from Earth, in the same way that houses on distant hills appear at a uniform distance. Stars in constellation patterns that appear to be related are in reality vast distances apart, often further from each other than they are from Earth. Astronomers measure the distance of stars by parallax (p.11) or by analysis of starlight, to find how far away they are from Earth or from each other.

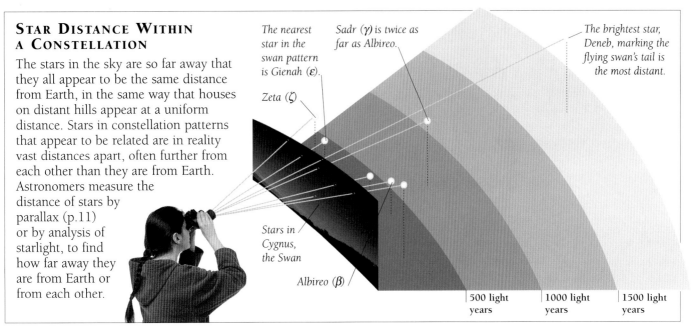

The nearest star in the swan pattern is Gienah (ε).

Sadr (γ) is twice as far as Albireo.

The brightest star, Deneb, marking the flying swan's tail is the most distant.

Zeta (ζ)

Stars in Cygnus, the Swan

Albireo (β)

| 500 light years | 1000 light years | 1500 light years |

# THE STARS

All stars are immense spinning globes of hot, luminous gas. They are born, live for millions of years, and eventually die. A star's mass, that is the amount of gas it is made of, has a direct bearing on its lifespan and the stages it will go through. All the stars we see in the night sky are of different ages and sizes; they vary in temperature, color, and luminosity. Some of them may not be single stars at all but double stars or a part of a star cluster (pp.86–7)

**The brightest star**
*The brightest star in the sky is Sirius, a white star. It shines at mag −1.46 and is visible from most countries of the world. Placed next to the Sun, it would outshine it 26-fold.*

**Beta (β)** *in Crux is a blue-white star of mag.1.25.*

**Acrux** *in Crux is a pair of stars of mag.0.83.*

**Arcturus** *in Boötes is orange-colored and the fourth brightest star in the sky at mag.−0.04.*

**Rigil Kentaurus** *in Centaurus is mag.−0.27, the third brightest star in the sky. It is actually a triple star.*

## THE APPEARANCE OF STARS

The stars we see in the sky are all at different distances from Earth, so their actual brightness cannot easily be compared. We can, however, compare how bright they appear in the sky. This is achieved using the apparent magnitude scale (p.16). In a constellation, the pattern is formed by the brightest stars (pp.82–3). The colors of stars is also observable from Earth (below)

## COLOR AND TEMPERATURE

At first sight, all stars seem to be the same color—brilliant white. After a while it becomes apparent that stars can be blue, white, yellow, orange, or red. It is possible to determine how hot a star is from its color. Blue stars are hottest, white stars are cooler, then come yellow, orange, and red.

*Deneb, the brightest star in the constellation, is a brilliant blue-white.*

*Omicron (o) is an orange and blue double star.*

*This star, epsilon (ε), is a yellow star.*

*Contrasting colors are easiest to see in stars that are close in the sky.*

**Observing star color**
*The various colors of the stars in the constellation Cygnus, the Swan, are easy to see in this photograph. Observing them in the real sky is not so easy but with practice you will be able to make them out.*

**Surface temperature**
*Stars are classified by astronomers according to their surface temperature. Below are the colors associated with the range of temperatures.*

90,000°F (50,000°C)

54,000°F (30,000°C)

18,000°F (10,000°C)

10,800°F (6,000°C)

7,200°F (4,000°C)

6,300°F (3,500°C)

## OUR STAR, THE SUN

The Sun is a yellow main-sequence star (see below), now in middle age. It will live for about 5 billion years longer. When its hydrogen is entirely used up, the Sun will become a red giant before beginning to die.

**Procyon** *in Canis Minor is mag.0.38.*

**Betelgeuse** *in Orion is a red star and mag.0.5.*

**Rigel** *in Orion is blue-white and the seventh brightest star in the sky.*

**Altair** *in Aquila forms a corner of a bright star triangle.*

**Vega** *in Lyra is the fifth brightest star in the sky.*

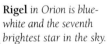

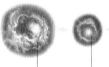
Sirius

## NOVAE

The brightness of most stars remains constant, or varies in a predictable way. A nova is a star that undergoes a sudden, unpredictable increase in brightness. This is because the star is one half of a binary system (p.86). Hydrogen transferred from its partner causes the brilliant outburst.

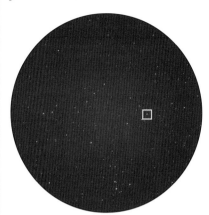

**Nova Cygni**
*In 1992 a star in Cygnus increased in brightness for several weeks before returning to its original magnitude.*

## THE SKY'S BRIGHTEST STARS

The 20 brightest stars in the sky are listed below in order of brilliance. All are easily seen with the naked eye.

| STAR NAME | CONSTELLATION | MAG. |
| --- | --- | --- |
| Sirius | CMa (p.113) | −1.46 |
| Canopus | Carina (p.103) | −0.72 |
| Rigil Kentaurus | Centaurus (p.102) | −0.27 |
| Arcturus | Boötes (p.120) | −0.04 |
| Vega | Lyra (p.125) | 0.03 |
| Capella | Auriga (p.112) | 0.08 |
| Rigel | Orion (p.112) | 0.12 |
| Procyon | CMi (p.110) | 0.38 |
| Achernar | Eridanus (p.132) | 0.46 |
| Betelgeuse | Orion (p.112) | 0.5 |
| Hadar | Centaurus (p.102) | 0.61 |
| Altair | Aquila (p.125) | 0.77 |
| Acrux | Crux (p.102) | 0.83 |
| Aldebaran | Taurus (p.107) | 0.85 |
| Antares | Scorpius (p.109) | 0.96 |
| Spica | Virgo (p.109) | 0.98 |
| Pollux | Gemini (p.106) | 1.14 |
| Fomalhaut | PsA (p.129) | 1.16 |
| Deneb | Cygnus (p.129) | 1.25 |
| Beta | Crux (p.102) | 1.25 |

## A STAR'S LIFE

Stars are born from a spinning gas and dust cloud. At the core, hydrogen gas is turned into helium and the star begins to shine. After millions of years the star starts to die.

Supergiant    Supernova

**Star sizes**
*A star alters size as it moves from stage to stage. A giant is at least 10 times bigger than the Sun, a supergiant about 100 times bigger.*

Nebula    Protostar    Main sequence star

**How stars are born**
*The spinning cloud of gas and dust (nebula) pulls material into the center (protostar) to shine as a main-sequence star.*

**How massive stars die**
*A massive star becomes very large and luminous (supergiant) before it explodes (supernova).*

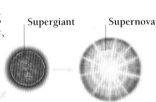
Red giant    Planetary nebula

**How stars like the Sun die**
*The star inflates and turns red (red giant). It blows off material as it dies (planetary nebula).*

The Sun
Red giant

# STAR FAMILIES

The majority of the stars we see in the night sky are single stars, but many are part of a double or multiple star system, with each unit of the team affecting the star's brightness, as we see it from Earth. All stars originate in a group or cluster of stars; we can see many of these in the sky. Tight-knit or globular clusters consist of old stars that have been together from birth and will stay together, but open clusters are groups of hot, young stars that will drift apart in time.

### DOUBLE STARS

Many stars are seen as double points of light as they have a companion. The two may be unrelated and the stars are merely optical doubles; others are physically related and close in space. They are called binaries (below).

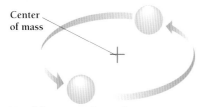

Center of mass

**Double stars of equal mass**
*In a double star system where both members have the same mass, the stars orbit a center positioned at a midway point between them.*

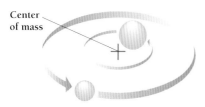

Center of mass

**Double stars of unequal mass**
*When one member of a double system is more massive than the other, the center of mass lies close to the heavier star.*

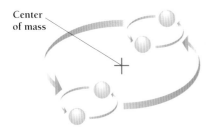

Center of mass

**A multiple star**
*In this drawing, there are four stars of equal mass forming two pairs, and both pairs of stars are orbiting a common center of mass.*

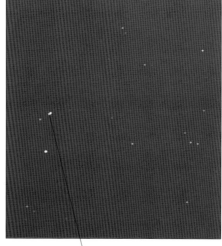

Alpha (α) in Capricornus

### LOOKING AT DOUBLE STARS

Many double stars can be seen with the naked eye or binoculars. Some stars are equal in brightness and color, others are not. A telescope is needed to separate close doubles.

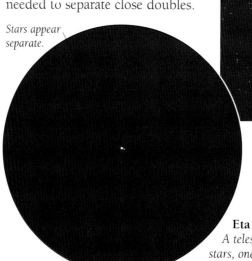

Stars appear separate.

Albireo B is a blue-green star.

Albireo A is a yellow star.

**Colored doubles**
*A single star, Albireo, marks the head of the swan in Cygnus. Binoculars show it to be two stars of contrasting color (pp.84–5). The brightest is golden yellow, while the second is blue-green.*

**Optical doubles**
*Binoculars show that Alpha (α) in Capricornus is two stars. They appear close together but only because they are on the same line of sight from Earth.*

Stars appear as one.

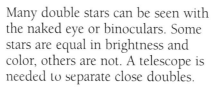

**Eta Cassiopeiae: naked-eye view**
*The bright star Eta Cassiopeiae is seen easily with the naked eye. It is just above the star Schedar, marking the second dip of the "W."*

**Eta Cassiopeiae: telescope view**
*A telescope will show Eta Cassiopeiae is two stars, one of mag.4 and one of mag.8. Their colors, yellow and red respectively, can just be seen.*

## STAR CLUSTERS

All stars are born in a cluster, but only some stay together for life. Open clusters of a few to a few thousand stars move apart. The tens to hundreds of thousands of stars in a globular cluster stay together.

**Open cluster**
*Stars in an open cluster are young and luminous. The clusters have no set structure, and are different in size and shape. Clusters such as the Double Cluster in Perseus (above) are best viewed with the naked eye or binoculars.*

**Globular cluster**
*A globular cluster is a swarm of old stars increasingly packed toward the center and usually spherical in shape. About 150 are known in our galaxy. Roughly half of them, including M10 in Ophiuchus (left), are visible through binoculars.*

## DOUBLE STARS

Use the planisphere to determine when double stars are high in the sky and hence in a good position for observing.

| NAME | CONSTELLATION |
| --- | --- |
| Mizar | Ursa Major (p.98) |
| Albireo | Cygnus (p.129) |
| Castor | Gemini (p.106) |
| Almach | Andromeda (p.132) |
| Algieba | Leo (p.106) |
| Cor Caroli | Canes Venatici (p.121) |
| Theta Orionis | Orion (p.112) |
| Epsilon Lyrae | Lyra (p.125) |

## STAR CLUSTERS

Star clusters are good objects to observe with binoculars. If you find it difficult to spot a cluster, use averted vision (p.15)

| NAME | CONSTELLATION |
| --- | --- |
| **Open clusters** | |
| Beehive, M44 | Cancer (p.106) |
| Double Cluster | Perseus (p.124) |
| Pleiades M45 | Taurus (p.107) |
| Jewel Box NGC4755 | Crux (p.102) |
| Wild Duck, M11 | Scutum (p.122) |
| M41 | Canis Major (p.113) |
| **Globular clusters** | |
| Omega Centauri | Centaurus (p.102) |
| 47 Iucanae | Tucana (p.103) |
| M13 | Hercules (p.124) |
| M22 | Sagittarius (p.108) |
| M15 | Pegasus (p.128) |
| M3 | Canes Venatici (p.121) |

## VARIABLE STARS

The brightness of some stars varies. This may be because a star is part of an eclipsing binary team (right), or because of a change in the star itself. Cepheid variables and Miras are in the latter category. Cepheids change over 1–50 days, Miras take longer.

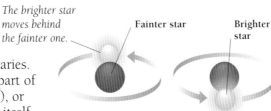

*The brighter star moves behind the fainter one.*

Fainter star

Brighter star

**Fainter**
*The brighter star is eclipsed by its fainter companion.*

**Brighter**
*The brighter star is now directly in view.*

**Variable star, Mira**
*Mira in the constellation Cetus is a red giant star that pulsates in size and flares significantly in brightness, from mag.10 to mag.3 in 332 days. It has given its name to the category of variable stars, which are known as Mira stars.*

Mira at mag.10          Mira at mag.3

## VARIABLE STARS

Familiarize yourself with the brightness of stars near a variable star and compare their relative brightnesses.

| NAME | CONSTELLATION |
| --- | --- |
| **Cepheids** | |
| Delta (δ) Cephei | Cepheus (p.98) |
| Eta (η) Aquilae | Aquila (p.125) |
| Zeta (ζ) Geminorum | Gemini (p.106) |
| Beta (β) Doradus | Dorado (p.103) |
| **Miras** | |
| Mira | Cetus (p.133) |
| Chi (χ) Cygni | Cygnus (p.129) |
| **Eclipsing binaries** | |
| Algol | Perseus (p.124) |
| Delta (δ) Librae | Libra (p.109) |

# NEBULAE

Nebulae (sing. nebula) are clouds of interstellar gas and dust. They are stages in star birth and death. When a massive star dies, it blows off gas and dust, and eventually this material coalesces to form new stars. Many nebulae can be seen with the naked eye; they look like fuzzy, light patches or dark, empty holes. They can be seen more clearly through binoculars or a telescope and make good subjects for photography and CCD imaging.

## TYPES OF NEBULA

Emission and reflection nebulae and supernova remnants are all bright nebulae. Dark nebulae look like dark patches in an otherwise bright sky.

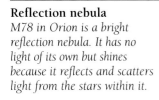

**Reflection nebula**
*M78 in Orion is a bright reflection nebula. It has no light of its own but shines because it reflects and scatters light from the stars within it.*

**Supernova remnant**
*This trail of gas, the Veil Nebula in Cygnus, is the remains of a supernova. The star that produced the material exploded about 50,000 years ago.*

*A dark lane of dust seems to cut the Great Nebula in two, but the two halves are part of one enormous cloud.*

**Dark nebula**
*This S-shaped cloud, known as the Snake Nebula, in Ophiuchus, is a dark nebula— a cloud of gas and dust dense enough to blot out the light of stars behind it.*

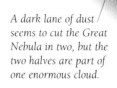

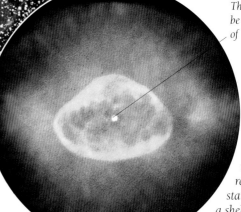

*The dying star can be seen at the center of this ring of gas.*

**Planetary nebula**
*A planetary nebula is the result of a dying star pushing off a shell of gas. This one (left) is the Blinking Planetary in Cygnus.*

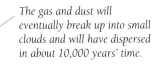

*The gas and dust will eventually break up into small clouds and will have dispersed in about 10,000 years' time.*

*Hot young stars illuminate the nebula. The four in the center are called the Trapezium.*

*The predominant red color of this emission nebula is produced by the interaction of gas particles in the nebula.*

## SUPERNOVAE

Every second, somewhere in the universe, a star explodes. These events are called supernovae. Most are out of sight but two or three per century have occurred in our own galaxy. The light from the explosion looks like an incredibly bright new star.

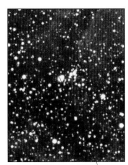

*Before it blew up, this blue supergiant looked like an ordinary star.*

*The star exploded on February 23, 1987 and is now supernova 1987A.*

### NEBULAE TO OBSERVE

Bright nebulae look like blurred, misty patches of light to the naked eye. You may need to use the technique of averted vision (p.15) to spot them.

| NAME | CONSTELLATION | TYPE |
|---|---|---|
| Orion, M42 | Orion (p.112) | Emission |
| Omega, M17 | Sagittarius (p.108) | Emission |
| Coalsack | Crux (p.102) | Dark |
| Cygnus Rift | Cygnus (p.129) | Dark |
| Merope | Taurus (p.107) | Reflection |
| Helix, NGC 7293 | Aquarius (p.108) | Planetary |
| Dumbbell, M27 | Vulpecula (p.129) | Planetary |
| Ring, M57 | Lyra (p.125) | Planetary |
| Veil, NGC6992 | Cygnus (p.129) | Supernova remnant |
| Crab, M1 | Taurus (p.107) | Supernova remnant |

## EMISSION NEBULA

The Great Nebula in Orion is one of the brightest emission nebulae in the sky. Light from the stars embedded in the nebula is absorbed by the hydrogen gas and re-emitted.

*The Trifid is an emission nebula.*

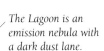

*The Lagoon is an emission nebula with a dark dust lane.*

## THE LAGOON (M8) AND TRIFID (M20) NEBULAE

This region of sky in Sagittarius is full of nebulous material. Two bright regions stand out. The bright cloud at the bottom is the Lagoon Nebula, M8, and the smaller cloud at the top is M20, the Trifid.

**Observing nebula**
*A telescope is a good way to see the details in nebulae, such as shape and color.*

# GALAXIES

Galaxies are vast collections of stars. Every galaxy consists of millions, or billions, of stars and there are thought to be billions of galaxies in the universe. The galaxy to which our Sun and Solar System belong is called the Milky Way. From Earth we can see other galaxies, but even the nearest are so remote they look like faint blurs of light to the naked eye or even through binoculars.

## THE SHAPE OF GALAXIES

Galaxies conform to four main shapes as shown below. The Milky Way is a spiral galaxy with a dense nucleus and spiral arms. Some galaxies are barred spirals and some are elliptical, while others have no definite shape and are called irregular.

*The central nucleus is packed with old stars.*

**Spiral galaxy**

**Barred spiral galaxy**

**Elliptical galaxy**

**Irregular galaxy**

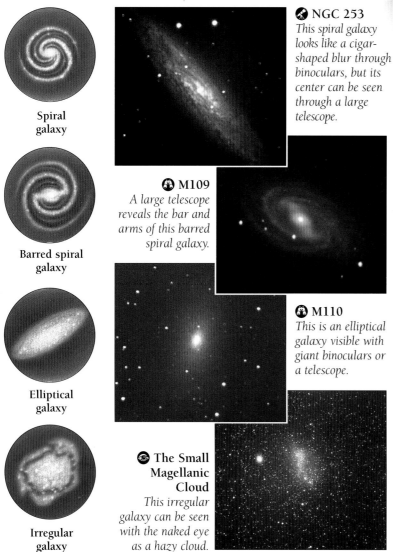

**● NGC 253**
*This spiral galaxy looks like a cigar-shaped blur through binoculars, but its center can be seen through a large telescope.*

**● M109**
*A large telescope reveals the bar and arms of this barred spiral galaxy.*

**● M110**
*This is an elliptical galaxy visible with giant binoculars or a telescope.*

**● The Small Magellanic Cloud**
*This irregular galaxy can be seen with the naked eye as a hazy cloud.*

## ANATOMY OF A GALAXY

This spectacular CCD image of galaxy NGC 2997 shows the magnificent spiral structure. In the center is a nucleus packed with old red stars, and in the spiral arms are the young, hot blue stars and new stars in the process of forming.

### GALAXIES TO OBSERVE

Use a known nearby star as a marker to locate a galaxy. If you still cannot easily see it in the sky, use averted vision (p.15); this will help you to spot it.

| NAME | CONSTELLATION | TYPE |
|---|---|---|
| ● Large Magellanic Cloud | Dorado (p.103) | Irregular |
| ● Small Magellanic Cloud | Tucana (p.103) | Irregular |
| ● Andromeda (M31) | Andromeda (p.132) | Spiral |
| ● Triangulum (M33) | Triangulum (p.133) | Spiral |
| ● Whirlpool (M51) | Canes Venatici (p.121) | Spiral |
| ● M95 | Leo (p.106) | Bar. Spiral |
| ● M49 | Virgo (p.109) | Elliptical |
| ● M87 | Virgo (p.109) | Elliptical |

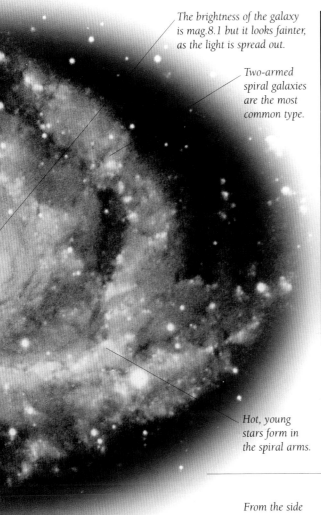

*The brightness of the galaxy is mag.8.1 but it looks fainter, as the light is spread out.*

*Two-armed spiral galaxies are the most common type.*

*Hot, young stars form in the spiral arms.*

## GALACTIC CLUSTERS

Galaxies exist together in clusters. The Fornax cluster (above) contains countless spiral, barred spiral, and elliptical galaxies. Our own galaxy, the Milky Way (below), is part of a cluster of about 30 galaxies known as the Local Group, which includes the Large and Small Magellanic Clouds (p.103) and the Andromeda Galaxy (p.132).

*From the side the spiral arms look like a flat disk.*

### Side view of our galaxy
*If we could see the Milky Way from the outside, and viewed edge-on, we would see a disk with a concentration of stars forming a central bulge.*

*The halo contains the oldest stars.*

**100,000 light years**

## THE MILKY WAY

Our galaxy, the Milky Way, is a spiral galaxy containing about 500,000 million stars. It is about 100,000 light years across. The Sun and Earth lie about two-thirds of the way from the center, on one of the spiral arms, the Orion arm. The Sun, and every other star in the Milky Way, follows its own orbit around the galactic center. The Sun takes about 220 million years to complete one orbit of the galaxy.

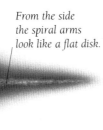

Orion arm

Sagittarius arm

Perseus arm

### Overhead view of our galaxy
*Seen face-on, the Milky Way's central bulge of stars is surrounded by spiral arms.*

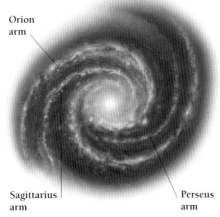

**The Milky Way viewed from Earth**
*The concentration of stars in our galaxy forms a river of light across the dark sky. The sight has been compared to spilled milk, hence the name of our galaxy, the Milky Way.*

# THE MILKY WAY

Apart from other galaxies, everything we see in the night sky—stars, star clusters, nebulae, the Sun, Moon, and planets—are part of our own galaxy, the Milky Way. From our position, two thirds of the way from the center of the galaxy (pp.90–1), we see most of the stars of the galaxy's disk as a river of light; this is commonly called the Milky Way. The density of the band of light varies, depending on the direction we look in. Toward the center, the band is denser, with more stars; away from the galactic center we see fewer stars. The map below shows the path of the Milky Way and the constellations through which it flows.

**The Milky Way in Aquila and Cygnus**
*For northern hemisphere observers, the Milky Way's path is brightest in Aquila and Cygnus. It can be seen with the naked eye, away from city lights; the darker the sky, the brighter the milky path.*

*The dark, naked-eye dust lane breaking the path in Cygnus is called the Cygnus Rift or Northern Coalsack.*

## LOCATING THE MILKY WAY

The path of the Milky Way can be seen from both the northern and southern hemispheres. Use the planisphere (pp.22–3) to check which part of its path is visible for your latitude and the time.

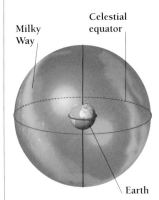

**The celestial sphere**
*The path of the Milky Way seems to be fixed to the sphere above our heads. It is shown here in pale blue.*

Celestial equator

Milky Way

Earth

Cassiopeia (p.99)

Cepheus (p.98)

Lacerta (p.129)

Cygnus (p.129)

Auriga (p.112)

Perseus (p.133)

Andromeda (p.132)

Taurus (p.107)

Vulpecula (p.129)

Sagitt. (p.124)

Aquil (p.12?)

*Cassiopeia marks the northerly turning point of the path of the Milky Way.*

*Turn binoculars to any patch of sky in the Milky Way's path to see many more stars than those visible to the naked eye.*

*Stars are particularly dense in the region of Sagittarius. When looking this way we are looking to the heart of the Galaxy.*

**The Milky Way surrounding Scutum**
*The tiny constellation Scutum is positioned at the center of this photograph, but its star pattern has been lost to view against the dense path of the Milky Way. This stretch of the Milky Way's path is easily seen with the naked eye.*

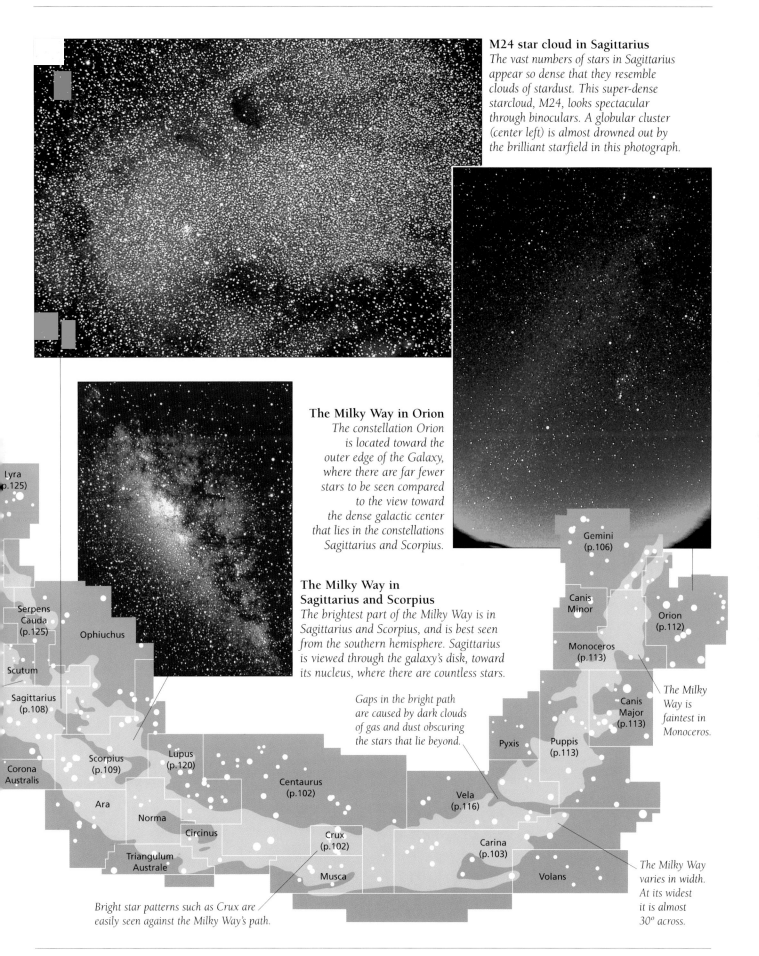

**M24 star cloud in Sagittarius**
*The vast numbers of stars in Sagittarius appear so dense that they resemble clouds of stardust. This super-dense starcloud, M24, looks spectacular through binoculars. A globular cluster (center left) is almost drowned out by the brilliant starfield in this photograph.*

**The Milky Way in Orion**
*The constellation Orion is located toward the outer edge of the Galaxy, where there are far fewer stars to be seen compared to the view toward the dense galactic center that lies in the constellations Sagittarius and Scorpius.*

**The Milky Way in Sagittarius and Scorpius**
*The brightest part of the Milky Way is in Sagittarius and Scorpius, and is best seen from the southern hemisphere. Sagittarius is viewed through the galaxy's disk, toward its nucleus, where there are countless stars.*

*Gaps in the bright path are caused by dark clouds of gas and dust obscuring the stars that lie beyond.*

*The Milky Way is faintest in Monoceros.*

*The Milky Way varies in width. At its widest it is almost 30° across.*

*Bright star patterns such as Crux are easily seen against the Milky Way's path.*

Lyra (p.125)

Serpens Cauda (p.125)

Ophiuchus

Scutum

Sagittarius (p.108)

Corona Australis

Scorpius (p.109)

Ara

Norma

Circinus

Lupus (p.120)

Triangulum Australe

Crux (p.102)

Musca

Centaurus (p.102)

Carina (p.103)

Volans

Vela (p.116)

Pyxis

Puppis (p.113)

Canis Major (p.113)

Monoceros (p.113)

Canis Minor

Gemini (p.106)

Orion (p.112)

# HOW TO USE THE SKY MAPS

The whole of the celestial sphere is shown as a series of sky maps on pp.96–133. These maps have been divided into polar maps and bimonthly maps, and have been designed to be used with the planisphere (pp.22–3). Each sky map is followed by clear, detailed maps of all the key constellations for that polar and bimonthly sky. The zodiacal constellations also appear on the monthly maps, but have more detailed treatment and maps on pp.106–9.

**North polar map**

**South polar map**

**Bimonthly map**

## DIVIDING THE CELESTIAL SPHERE

For ease of reference, the sphere has been divided into top and bottom sections showing the north and south polar sky, and six vertical segments showing the bimonthly sky.

## THE BIMONTHLY MAPS

Together with the polar maps, these six bimonthly maps show all the constellations visible in the sky throughout the year. Because the night sky seems to move from east to west as seen from Earth, the maps are used from right to left, starting in January and ending in December.

**November and December**

**September and October**

**July and August**

## THE ZODIAC

This map shows the zodiacal constellations in relation to the celestial equator and the ecliptic (the path of the Sun across the sky as seen from Earth). It is followed by a detailed map for each zodiacal constellation. This same map is also used for locating the planets (pp.40–61).

*The south celestial pole has no convenient star to act as a marker.*

*The Sun, Moon, and all planets move within the band of the zodiac.*

**South Polar Stars**

## THE SOUTH POLAR MAP

This map is centered on the south celestial pole showing constellations between 50°S and 90°S. Some of these stars never set for some observers, being "circumpolar." The map (pp.100–101) is followed by constellation maps.

**The Zodiac**

*The Milky Way is shown in pale blue on the star maps.*

*The celestial equator is a projection of the Earth's equator onto the celestial sphere.*

## THE NORTH POLAR MAP

This map shows the constellations between 50°N and 90°N. Some of the stars in these constellations never set for some observers and are "circumpolar." This map (pp.96–7) is followed by constellation maps.

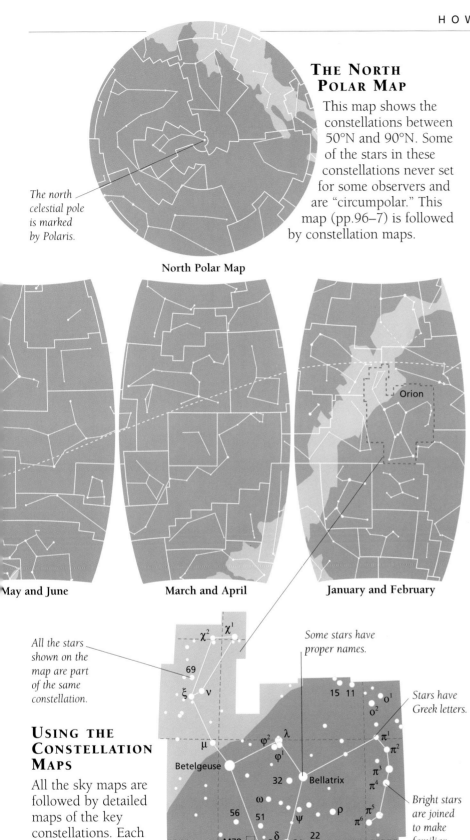

*The north celestial pole is marked by Polaris.*

**North Polar Map**

**May and June**

**March and April**

**January and February**

Orion

## USING THE CONSTELLATION MAPS

All the sky maps are followed by detailed maps of the key constellations. Each of these maps show the constellation shape, all the stars in a constellation that can be seen with the naked eye, and a selection of interesting sights for binoculars and telescopes.

*All the stars shown on the map are part of the same constellation.*

*Some stars have proper names.*

*Stars have Greek letters.*

*Bright stars are joined to make familiar shapes.*

*Symbols denote deep-sky objects of particular interest.*

χ² χ¹
69
ξ ν
μ
φ² λ
φ¹
Betelgeuse
32 Bellatrix
ω ρ
56 51 ψ
M78 ε δ 31 22
Horsehead nebula ζ
σ η
1981
M42
49 υ τ
29
Saiph Rigel

15 11 o¹
o²
π¹
π²
π³
π⁴
π⁵
π⁶

## KEY TO THE CONSTELLATION MAPS

The constellation maps are designed to be used with the sky maps and planisphere (pp.22–3). When you have identified the part of the sky you wish to view, use the constellation map for that area which gives details of the bright stars, the constellation shapes, and any deep-sky objects that can be easily observed.

### Deep-sky objects

*Symbols are used to represent different types of deep-sky objects. Often, a number or name also appears next to an object. These are official classification numbers or names.*

 Galaxy

 Globular Cluster

 Open Cluster

 Diffuse Nebula

 Planetary Nebula

### Measuring the constellations

*The sizes of constellations can be difficult to gauge when you are looking at the sky, and so a hand measurement for the width of each constellation has been given. (See also Measuring the Sky, p.21).*

1 hand    ½ hand    = 2½ hands

### Naming the stars

*Stars within a constellation are identified by a Greek letter. The star marked Alpha (α) is usually the brightest. Many bright stars also have a proper name.*

**The Greek Alphabet**

| | | | |
|---|---|---|---|
| α – Alpha | η – Eta | ν – Nu | τ – Tau |
| β – Beta | θ – Theta | ξ – Xi | υ – Upsilon |
| γ – Gamma | ι – Iota | ο – Omicron | φ – Phi |
| δ – Delta | κ – Kappa | π – Pi | χ – Chi |
| ε – Epsilon | λ – Lambda | ρ – Rho | ψ – Psi |
| ζ – Zeta | μ – Mu | σ – Sigma | ω – Omega |

### Star magnitude

*Each star is represented on the map by a white dot. The size of dot indicates a star's magnitude, its apparent brightness when seen from the Earth. The brightest star has the largest dot, but the smallest number.*

**Magnitude scale**

-1  0  1  2  3  4  5  6

# NORTH POLAR SKY

Some constellations are in the sky at all times for northern observers, depending on latitude. They do not set below the horizon and seem to trace out a circular path around a single point in the sky. This point is known as the north celestial pole and the apparent circular path of the stars is due to the daily rotation of Earth. These stars are circumpolar, and they complete one circle of the sky in just under 24 hours. (Key constellations are on pp.98–9.)

*The faint path of the Milky Way can only be seen on very dark nights.*

Lacerta

M52

Cygnus

Cepheus

## LOCATING THE CONSTELLATIONS

The north celestial pole is at the center of this map. Observers at the North Pole on Earth would see this point directly overhead, and the stars around it would always be visible. More southerly observers would see the north celestial pole at a lower point in their sky. Set the planisphere to discover which stars are circumpolar for you.

*Cepheus has several variable stars (p.87).*

Draco

Ursa Minor

*The entire constellation Ursa Minor is visible to observers north of the equator on Earth.*

Mizar

Alcor

### 🔭 Open star cluster M52 in Cassiopeia
*There are about 100 stars packed together in open cluster M52. Observers can find the cluster by first locating Cassiopeia's bright stars, Alpha (α) and Beta (β) and then extending a line from them. The cluster can be seen with binoculars but its brightest stars (mag.9–10) can only be seen with a telescope. M52 lies 5,200 light years away from Earth.*

### 🔭 Mizar and Alcor in Ursa Major
*The bright star Mizar in the tail of Ursa Major has a companion called Alcor. Both can be seen with binoculars or by observers with keen eyesight. A telescope will reveal that Mizar itself is a double star, consisting of Mizar A and Mizar B.*

| KEY TO DEEP-SKY OBJECTS | The Milky Way | Galaxy | Globular Cluster | Open Cluster | Diffuse Nebula | Planetary Nebula |
| --- | --- | --- | --- | --- | --- | --- |

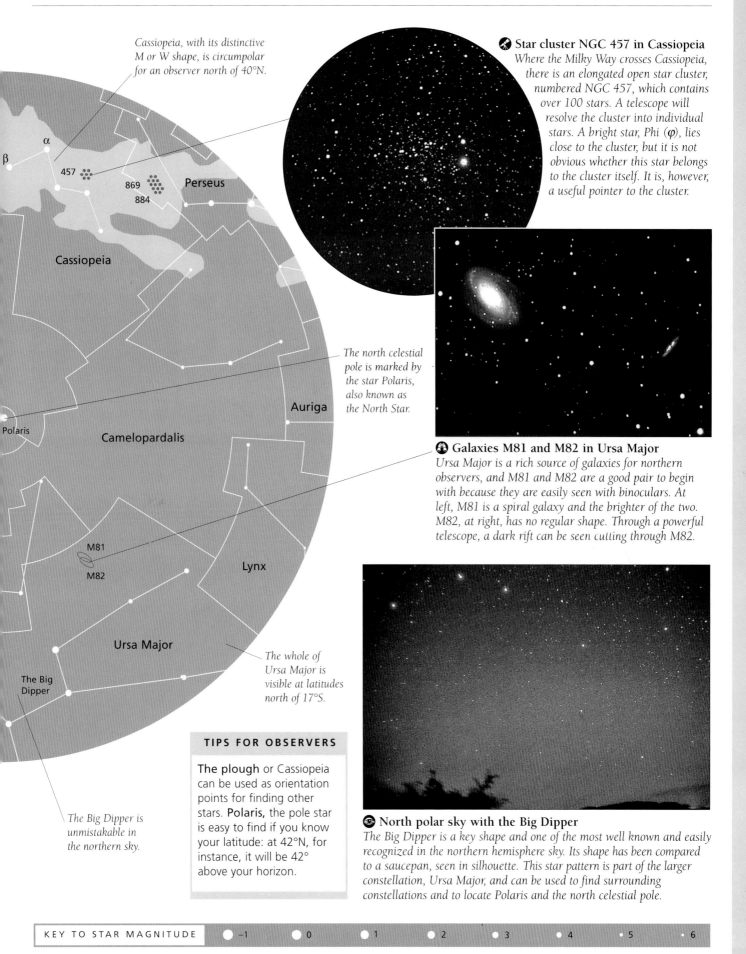

*Cassiopeia, with its distinctive M or W shape, is circumpolar for an observer north of 40°N.*

**Star cluster NGC 457 in Cassiopeia**
*Where the Milky Way crosses Cassiopeia, there is an elongated open star cluster, numbered NGC 457, which contains over 100 stars. A telescope will resolve the cluster into individual stars. A bright star, Phi (φ), lies close to the cluster, but it is not obvious whether this star belongs to the cluster itself. It is, however, a useful pointer to the cluster.*

*The north celestial pole is marked by the star Polaris, also known as the North Star.*

**Galaxies M81 and M82 in Ursa Major**
*Ursa Major is a rich source of galaxies for northern observers, and M81 and M82 are a good pair to begin with because they are easily seen with binoculars. At left, M81 is a spiral galaxy and the brighter of the two. M82, at right, has no regular shape. Through a powerful telescope, a dark rift can be seen cutting through M82.*

*The whole of Ursa Major is visible at latitudes north of 17°S.*

### TIPS FOR OBSERVERS

**The plough** or Cassiopeia can be used as orientation points for finding other stars. **Polaris**, the pole star is easy to find if you know your latitude: at 42°N, for instance, it will be 42° above your horizon.

*The Big Dipper is unmistakable in the northern sky.*

**North polar sky with the Big Dipper**
*The Big Dipper is a key shape and one of the most well known and easily recognized in the northern hemisphere sky. Its shape has been compared to a saucepan, seen in silhouette. This star pattern is part of the larger constellation, Ursa Major, and can be used to find surrounding constellations and to locate Polaris and the north celestial pole.*

KEY TO STAR MAGNITUDE    ● −1    ○ 0    ● 1    ● 2    ● 3    • 4    · 5    6

# NORTH POLAR SKY: Key Constellations

Ursa Major and Ursa Minor are useful signposts in the north polar sky. Once found, they help locate the celestial pole, which is marked by Polaris, the North Star. As the Earth turns, they circle the sky, standing above the pole, then hanging below it.

## URSA MINOR *The Little Bear*

Width

At the tip of the bear's tail is Polaris, the North Star. This star lies close to the north celestial pole and appears to remain stationary as the other stars move around it. Polaris is a Cepheid variable star (p.87), that alters in brightness every four days.

Polaris

δ

*Polaris marks the north celestial pole.*

ε

ζ

4

5

η

Kochab

Pherkad

👁 **Polaris**
*All the stars of Ursa Minor, including Polaris, can be easily seen with the naked eye.*

## CEPHEUS *Cepheus*

Width

Cepheus and Cassiopeia, representing the King and Queen of Ethiopia, lie together in the sky. Cepheus contains star clusters and noted stars such as the red Garnet Star (μ) and Delta (δ). Delta was the first Cepheid variable to be discovered, and gave its name to the entire category.

γ

κ

π

24

11

β

o

ι

θ

ξ

α

η

9

ν

δ

ζ

μ

ε

👁 **Red Garnet Star**
*The red Garnet Star (μ) is easy to spot below the main shape of Cepheus.*

*Delta (δ) is a yellow supergiant, with a blue-white companion.*

## URSA MAJOR *The Great Bear*

Width

Seven bright stars in Ursa Major form the Big Dipper, one of the most familiar shapes in the northern sky. The entire constellation, however, is much bigger, with the Big Dipper marking just the hindquarters and tail of the bear. Mizar in the tail is a multiple star (p.86). Among the stars near the bear's head is a pair of galaxies, M81 and M82.

M82

24

ρ

M81

σ

π²

τ

o

23

υ

M101

α

83

Alcor

80

78

δ

Plough

β

36

18

Mizar

ε

φ

26

θ

15

η

M108

γ

M109

M97

ι

κ

χ

ψ

M81 *is a spiral galaxy and M82 is an irregular galaxy.*

56

ω

μ

λ

*Mizar (mag.2) is a multiple star with a close companion, Alcor (mag.4).*

55

ν

ξ

*Galaxy M108 appears as a faint blur through binoculars.*

🔭 **Spiral galaxy M81**
*Seen through a telescope, this galaxy appears as an oval with a bright center. If conditions are good, M81 can be picked out with the naked eye.*

| KEY TO DEEP-SKY OBJECTS | The Milky Way | Galaxy | Globular Cluster | Open Cluster | Diffuse Nebula | Planetary Nebula |

## DRACO *The Dragon*

Width

Draco is not an easy constellation to make out. It is one of the largest constellations in the sky and winds around Ursa Minor. To find the dragon, start by locating the four stars that mark the dragon's head. Imagination is needed to trace out the rest. This constellation is noted for double stars. Nu (ν), the faintest star in the dragon's head, is a wide double star that can be seen through binoculars.

*Ursa Minor is almost surrounded by Draco.*

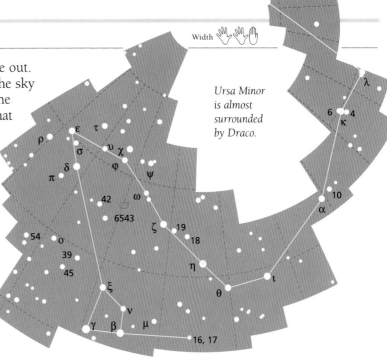

👁 **Stars in Draco**
*The most prominent stars in this constellation are in the dragon's head. The others are not easy to pick out with the eye alone.*

## CAMELOPARDALIS *The Giraffe*

Width

To find this constellation look between the constellation Auriga (p.112) and the star Polaris (see opposite). Camelopardalis covers a large area of sky but has few objects of interest. The brightest star Beta (β) is only mag.4. This star can be used to find open star cluster NGC 1502, which can be seen with binoculars as a patch of light.

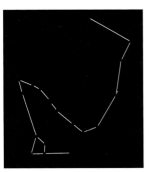

🔭 **Kemble's Cascade**
*This string of about 25 stars lies close to the star cluster NGC 1502. Only the brightest of the stars can be seen with the naked eye.*

## CASSIOPEIA *Cassiopeia*

Width

The bright stars in Cassiopeia make a recognizable shape in the sky. When the stars are below the north celestial pole they form a "W" and when above it they form an "M." The most interesting star is Gamma (γ), a blue giant that alters in brightness as rings of gas are blown off. Cassiopeia also has several notable star clusters, such as open cluster M52 and NGC 457, an elongated open cluster.

*M103 is a star cluster, visible with giant binoculars.*

*NGC 457 is fine open star cluster.*

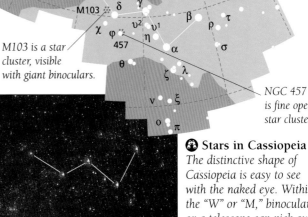

👁 **Stars in Cassiopeia**
*The distinctive shape of Cassiopeia is easy to see with the naked eye. Within the "W" or "M," binoculars or a telescope can pick out scores of stellar objects.*

| KEY TO STAR MAGNITUDE | | –1 | 0 | 1 | 2 | 3 | 4 | 5 | 6 |
|---|---|---|---|---|---|---|---|---|---|

# SOUTH POLAR SKY

Some constellations are in the sky at all times for southern observers, depending on latitude. They do not set below the horizon, but seem to trace out a circular path around a single point. This point is known as the south celestial pole and the stars around it are circumpolar, completing one circle of sky in under 24 hours. (Key constellations are on pp.102–3.)

## LOCATING THE CONSTELLATIONS

The south celestial pole is at the center of this map. Observers at the South Pole on Earth would see this point directly overhead and the stars around it would always be visible. More northerly observers would see the celestial pole at a lower point in their sky. Set the planisphere to see which stars are circumpolar for you.

*Tucana is only completely visible to observers located south of 14°N.*

*This hazy patch of light is the Small Magellanic Cloud. It is circumpolar for observers south of 20°S.*

*Dorado is completely visible for observers south of 20°N.*

*The Large Magellanic Cloud is always above the horizon for observers south of 25°S.*

Phoenix
Achernar
Horologium
Reticulum
47 Tucanae
Small Magellanic Cloud
Hydrus
Dorado
Pictor
Large Magellanic Cloud
Tarantula Nebula
Mensa
Chameleon
Volans
Puppis
Carina
3372
3532
Vela

**Dorado and Tucana**
*The constellations Dorado and Tucana are easy to find in the south polar sky, because they each contain a galaxy that is visible to the naked eye. These galaxies are the Large and Small Magellanic Clouds, named after the explorer Ferdinand Magellan. The large one at the top of the picture is in Dorado and the small one is in Tucana.*

**Globular star cluster in Tucana**
*After Omega Centauri, 47 Tucanae is the most spectacular globular cluster in the southern sky. It lies very close to the Small Magellanic Cloud in Tucana and can be seen with the naked eye and binoculars, but a small telescope transforms it into a truly breathtaking sight.*

| KEY TO DEEP-SKY OBJECTS | The Milky Way | Galaxy | Globular Cluster | Open Cluster | Diffuse Nebula | Planetary Nebula |

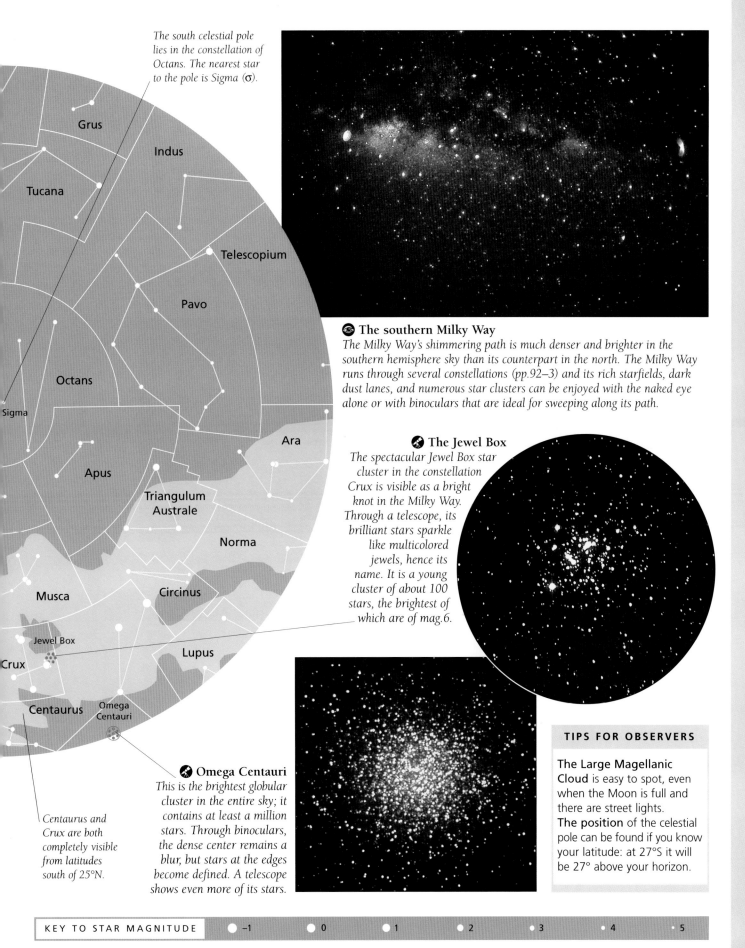

*The south celestial pole lies in the constellation of Octans. The nearest star to the pole is Sigma (σ).*

Grus

Indus

Tucana

Telescopium

Pavo

Octans

Sigma

Ara

Apus

Triangulum
Australe

Norma

Musca

Circinus

Jewel Box

Crux

Lupus

Centaurus

Omega
Centauri

*Centaurus and Crux are both completely visible from latitudes south of 25°N.*

### ◉ The southern Milky Way
*The Milky Way's shimmering path is much denser and brighter in the southern hemisphere sky than its counterpart in the north. The Milky Way runs through several constellations (pp.92–3) and its rich starfields, dark dust lanes, and numerous star clusters can be enjoyed with the naked eye alone or with binoculars that are ideal for sweeping along its path.*

### ✦ The Jewel Box
*The spectacular Jewel Box star cluster in the constellation Crux is visible as a bright knot in the Milky Way. Through a telescope, its brilliant stars sparkle like multicolored jewels, hence its name. It is a young cluster of about 100 stars, the brightest of which are of mag.6.*

### ✦ Omega Centauri
*This is the brightest globular cluster in the entire sky; it contains at least a million stars. Through binoculars, the dense center remains a blur, but stars at the edges become defined. A telescope shows even more of its stars.*

#### TIPS FOR OBSERVERS

**The Large Magellanic Cloud** is easy to spot, even when the Moon is full and there are street lights.
**The position** of the celestial pole can be found if you know your latitude: at 27°S it will be 27° above your horizon.

| KEY TO STAR MAGNITUDE | ● −1 | ● 0 | ● 1 | ● 2 | ● 3 | • 4 | · 5 |

# SOUTH POLAR SKY: Key Constellations

The constellations nearest to the south celestial pole have some of the finest sights in the night sky. The predominant constellations, Crux and Centaurus, are traversed by the Milky Way, and both constellations have bright stars and star clusters. Even the relatively unspectacular Dorado and Tucana contain the two Magellanic Clouds, companion galaxies to our own Milky Way.

## CRUX *The Southern Cross* Width 🖐

This may be the smallest constellation in the sky but it has a wealth of interesting objects and is very easy to find. Its four brilliant stars mark the shape of a cross and stand out clearly against the path of the Milky Way. A dark nebula (p.88), named the Coalsack, is silhouetted against the Milky Way, blocking out its light. Between the Coalsack and Beta (β) is the Jewel Box (NGC 4755), a magnificent open star cluster that is visible to the naked eye.

*The Jewel Box is a sparkling open star cluster.*

### ◉ Stars in Crux
*Of the four naked-eye stars that define the shape of the cross, three are white and one, Gamma (γ), is orange-red. Close to Beta (β) is the Jewel Box. Its brightest star, Kappa (κ), is visible to the naked eye.*

## CENTAURUS *The Centaur* Width 🖐🖐🖐

Like Crux, the large constellation Centaurus lies in the Milky Way and is packed with interesting sights. It contains two bright stars, Rigil Kentaurus and Hadar. These stars mark the front limbs of the centaur. Near Rigil Kentaurus is an 11th-magnitude star, called Proxima (not shown). Lying less that 4.3 light years away, Proxima is the nearest star to Earth. Near the center is Omega Centauri, the most spectacular globular cluster in the night sky. Centaurus also has some other fine star clusters, a planetary nebula, and an unusual galaxy, NGC 5128, or Centaurus A, which emits a strong radio signal.

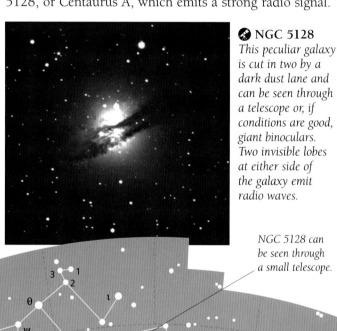

### ⊗ NGC 5128
*This peculiar galaxy is cut in two by a dark dust lane and can be seen through a telescope or, if conditions are good, giant binoculars. Two invisible lobes at either side of the galaxy emit radio waves.*

*NGC 5128 can be seen through a small telescope.*

*Omega Centauri appears as a fuzzy area larger than the Full Moon when seen with the naked eye or through binoculars.*

*Rigil Kentaurus, the third-brightest star in the night sky (mag. –0.3), can be seen with a telescope as a binary.*

*Hadar is a blue-white giant star of mag.0.6.*

*NGC 3918 is a planetary nebula that appears blue through a telescope.*

| KEY TO DEEP-SKY OBJECTS | The Milky Way | Galaxy | Globular Cluster | Open Cluster | Diffuse Nebula | Planetary Nebula |

## CARINA *The Keel*
Width ✋✋

At the far right of this constellation is the brilliant star Canopus, but it is the left-hand side, where the path of the Milky Way crosses the constellation, that is rich in stars, star clusters, and nebulae. Eta (η) can be seen with the naked eye. It is an unstable supergiant, 100 times the mass of our Sun, that lies in the center of the diffuse nebula NGC 3372.

*Canopus (mag.–1) is the second-brightest star in the night sky.*

### 🜨 Stars in Carina
*Carina is a good naked-eye and binocular constellation with numerous star clusters.*

## DORADO *The Goldfish*
Width ✋

Dorado contains the bulk of the Large Magellanic Cloud (LMC), our nearest galaxy. This galaxy can be seen with the naked eye as a light patch. Inside the patch, about the size of the Full Moon, is a fuzzy object, the Tarantula Nebula (NGC 2070), which binoculars show as a cloud of gas with intertwined rifts.

*Beta (β) is a Cepheid variable (p.87).*

### 🜨 The Large Magellanic Cloud
*Binoculars or a telescope show the irregular shape of this galaxy and hundreds of its stars.*

## OCTANS *The Octant*
Width ✋

Most stars in Octans are faint, and not easily visible to the naked eye. It is in this constellation that the south celestial pole lies. The nearest star to the south celestial pole is Sigma (σ). This constellation also contains Melotte 227, the southern-most visible open star cluster.

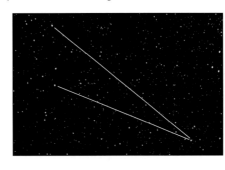

*Sigma (σ) is near to the south celestial pole.*

### 🜨 Stars in Octans
*The brightest stars make the triangular shape of the octant. None are brighter than mag.4.*

## TUCANA *The Toucan*
Width ✋✋

The constellation Tucana contains the Small Magellanic Cloud (SMC) and 47 Tucanae, two fascinating objects lying close together in the corner nearest to the south celestial pole. SMC can be seen with the naked eye but binoculars or a telescope will show some of its clusters and nebulae. 47 Tucanae is a globular cluster with tens of thousands of stars.

*Beta (β) is a star whose multiple parts can be seen with a telescope.*

*47 Tucanae is one of the finest star clusters in the sky.*

### 🜨 The Small Magellanic Cloud
*Shaped like a comma, this is the smaller of the Milky Way's two satellite galaxies. The other is its companion galaxy, the Large Magellanic Cloud.*

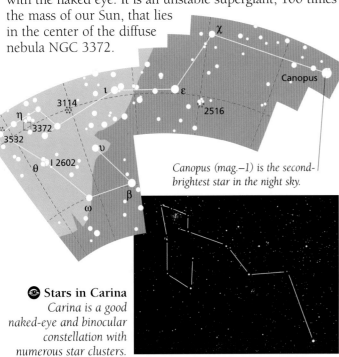

| KEY TO STAR MAGNITUDE | | ● –1 | ● 0 | ● 1 | ● 2 | ● 3 | • 4 | • 5 | • 6 |
|---|---|---|---|---|---|---|---|---|---|

# THE PATH OF THE ZODIAC

Twelve constellations have a particular significance for astronomers, because they form the backdrop to the ecliptic, the Sun's apparent yearly path through the stars (p.38). Collectively, these constellations are known as the zodiac. The Sun takes approximately one month to move through each constellation, taking a year to complete the circle. The path of the Sun—the ecliptic—is shown on the map below. On the pages that follow, there is a map for each of the constellations, and the most interesting features are described in detail.

## LOCATING THE CONSTELLATIONS

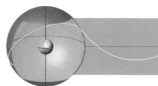

All observers can see the band of the zodiac, but northern observers will find it difficult to see the southernmost constellations, such as Scorpius, and southern observers will find it difficult to see the northernmost constellations, such as Gemini. Use the planisphere to find what is in the sky for your particular latitude.

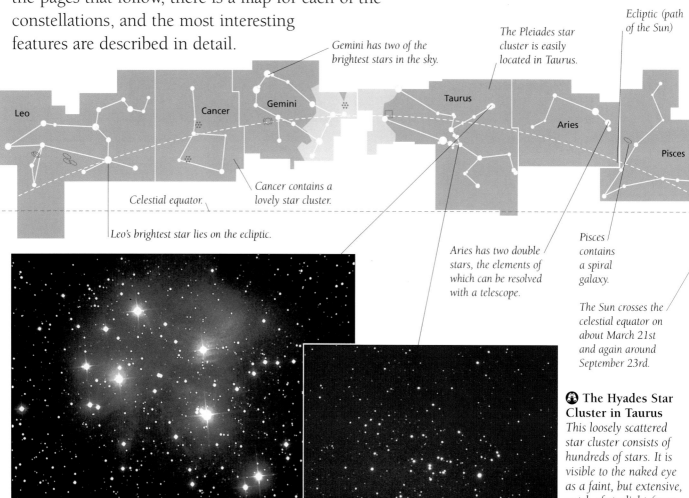

*Gemini has two of the brightest stars in the sky.*

*The Pleiades star cluster is easily located in Taurus.*

*Ecliptic (path of the Sun)*

Leo

Cancer

Gemini

Taurus

Aries

Pisces

*Cancer contains a lovely star cluster.*

*Celestial equator.*

*Leo's brightest star lies on the ecliptic.*

*Aries has two double stars, the elements of which can be resolved with a telescope.*

*Pisces contains a spiral galaxy.*

*The Sun crosses the celestial equator on about March 21st and again around September 23rd.*

**The Pleiades Star Cluster in Taurus**
*The alternative name for this star cluster, the "Seven Sisters," comes from the six or seven stars that can be seen with the naked eye. More of the stars in the cluster can be seen with binoculars and a telescope. This long-exposure photograph (above) also reveals the gas and dust cloud that produced the cluster.*

**The Hyades Star Cluster in Taurus**
*This loosely scattered star cluster consists of hundreds of stars. It is visible to the naked eye as a faint, but extensive, patch of starlight (mag. 0.5), with just a handful of stars shining brightly. Binoculars, with their wide field of view, show the group to best advantage because such an extensive area of sky is covered by the group.*

| KEY TO DEEP-SKY OBJECTS | The Milky Way | Galaxy | Globular Cluster | Open Cluster | Diffuse Nebula | Planetary Nebula |
|---|---|---|---|---|---|---|

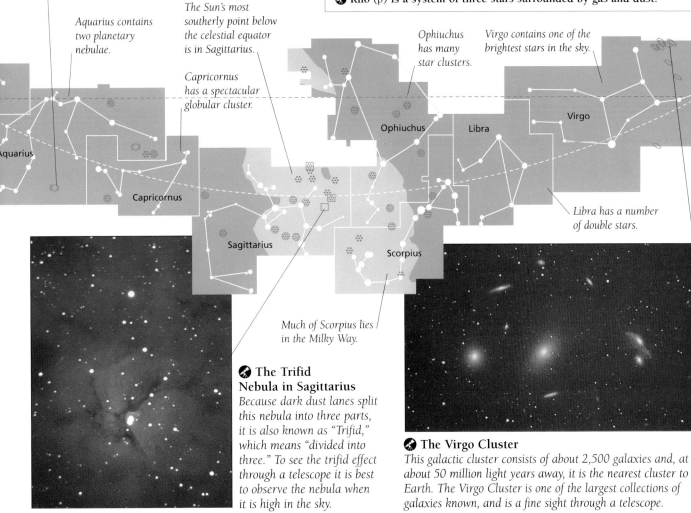

### ⊕ The Helix Nebula in Aquarius
*When viewed through binoculars, the Helix Nebula looks round, but through a telescope and in long-exposure photographs, its spectacular and distinctive shape becomes evident. This nebula is the remains of the material ejected by a dying star (pp.88–9).*

## OPHIUCHUS, 13TH ZODIACAL CONSTELLATION
Although traditionally there are considered to be only twelve zodiacal constellations, the Sun does move through the southern end of the constellation Ophiuchus, the Serpent Bearer, which qualifies this constellation as part of the zodiac. Ophiuchus is a large constellation with many faint and scattered stars and is located north of Scorpius and south of Hercules.

⊕ **Rho (ρ) is a system of three stars surrounded by gas and dust.**

Aquarius contains two planetary nebulae.

The Sun's most southerly point below the celestial equator is in Sagittarius.

Capricornus has a spectacular globular cluster.

Ophiuchus has many star clusters.

Virgo contains one of the brightest stars in the sky.

Aquarius

Capricornus

Sagittarius

Ophiuchus

Libra

Virgo

Scorpius

Libra has a number of double stars.

Much of Scorpius lies in the Milky Way.

### ⊕ The Trifid Nebula in Sagittarius
*Because dark dust lanes split this nebula into three parts, it is also known as "Trifid," which means "divided into three." To see the trifid effect through a telescope it is best to observe the nebula when it is high in the sky.*

### ⊕ The Virgo Cluster
*This galactic cluster consists of about 2,500 galaxies and, at about 50 million light years away, it is the nearest cluster to Earth. The Virgo Cluster is one of the largest collections of galaxies known, and is a fine sight through a telescope.*

| KEY TO STAR MAGNITUDE | –1 | 0 | 1 | 2 | 3 | 4 | 5 |
|---|---|---|---|---|---|---|---|

# ZODIACAL CONSTELLATIONS 1

Leo and Taurus are splendid constellations with bright stars and distinctive shapes. Taurus is prominent in the northern hemisphere sky in winter, and has two remarkable star clusters, the Hyades and the Pleiades. Next to Taurus is Gemini. This constellation marks the Sun's most northerly point in its yearly path. (All the constellations here are shown on the sky maps on pp.110–31.)

## LEO *The Lion*

Width

The stars in Leo really do make the shape of a lion; the bright stars of the lion's body and the curving head and neck can easily be traced in the sky. Below the lion's belly are five galaxies that are visible with binoculars. Leo is the constellation from which the Leonid meteor shower radiates annually in November (pp.74–5).

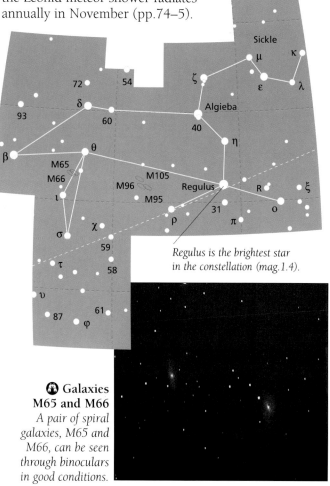

*Regulus is the brightest star in the constellation (mag.1.4).*

**☉ Galaxies M65 and M66**
*A pair of spiral galaxies, M65 and M66, can be seen through binoculars in good conditions.*

## CANCER *The Crab*

Width

The constellation Cancer is the least impressive of the zodiacal constellations as its brightest stars are relatively faint (mag.4). To locate the constellation, look between the bright stars of Leo and Gemini. In the center of Cancer is Praesepe (M44), a beautiful open cluster that is commonly known as the Beehive Cluster.

**☉ The Beehive Cluster**
*Through binoculars, this star cluster looks like a swarm of bees, hence its name.*

## GEMINI *The Twins*

Width

The two bright stars Castor (mag.1.6) and Pollux (mag.1.2), that mark the head of the twins in Gemini, are unmistakable in the northern hemisphere winter sky. It is from this part of the sky, near Castor, that the Geminid meteor shower (pp.74–5) radiates annually in December.

*This rich open cluster is just visible to the naked eye.*

**☉ The Clownface Nebula**
*Viewed through a small or large telescope, planetary nebula NGC 2392 looks like a bluish disk.*

---

| KEY TO DEEP-SKY OBJECTS | The Milky Way | Galaxy | Globular Cluster | Open Cluster | Diffuse Nebula | Planetary Nebula |
|---|---|---|---|---|---|---|

## TAURUS *The Bull*

Width

Taurus is a fine constellation for naked-eye and binocular observers alike. The brightest star, Aldebaran (mag.0.85), is a red giant and sometimes called "the eye of the bull." Two of the best star clusters in the sky are located in Taurus. One, the Pleiades, or Seven Sisters, contains about 100 stars, including a number of double stars. The other is the Hyades, a large cluster forming the "V" shape of the bull's face.

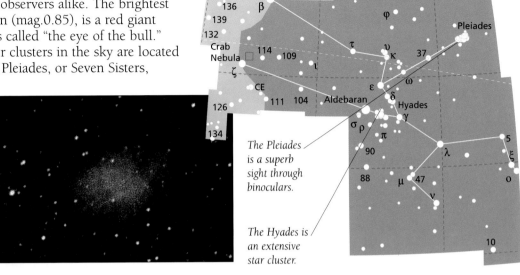

### 🌀 The Crab Nebula
*This is the remains of a supernova that occurred in AD 1054. The crab-like shape can be seen with a telescope.*

*The Pleiades is a superb sight through binoculars.*

*The Hyades is an extensive star cluster.*

## ARIES *The Ram*

Width

Compared to its neighbor, Taurus, Aries is a disappointing constellation. It has only one bright star, Hamal (Arabic for sheep) which is mag.2. This star can be found by first locating the Pleiades star cluster (see above) and then moving to the right. In the ram's head are many stars that are fainter than Hamal, two of which are double stars. Their companion stars can be seen through binoculars or a small telescope.

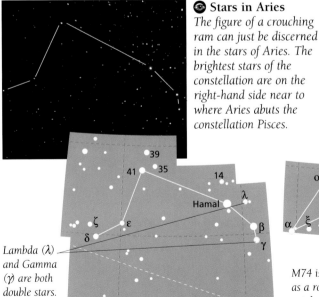

### 👁 Stars in Aries
*The figure of a crouching ram can just be discerned in the stars of Aries. The brightest stars of the constellation are on the right-hand side near to where Aries abuts the constellation Pisces.*

*Lambda (λ) and Gamma (γ) are both double stars.*

## PISCES *The Fishes*

Width

Pisces represents two fishes joined together by ribbon. It is an unremarkable constellation as it contains no very bright stars, but it does have an interesting asterism (star group). This is an elliptical configuration of stars known as the Circlet, that can be seen just south of the Great Square of Pegasus (p.128).

### 🌀 Galaxy M74
*This faint spiral faces Earth and can be seen only with a large telescope or with CCD equipment.*

*M74 is visible only as a rounded misty patch through a small telescope.*

*The Circlet is easy to pick out because the area is lacking in stars.*

| KEY TO STAR MAGNITUDE | ● −1 | ● 0 | ● 1 | ● 2 | ● 3 | ● 4 | • 5 | · 6 |

# ZODIACAL CONSTELLATIONS 2

Sagittarius, Scorpius, and Virgo are the most impressive of constellations; they have bright stars and deep-sky objects for naked-eye, binocular, and telescope observers alike. (The constellations here are shown on the sky maps on pp.110–31.)

## AQUARIUS *The Water Carrier*    Width

Aquarius has some fine planetary nebulae (p.88) and globular clusters (p.87). The Helix Nebula (NGC 7293) can be seen as a faint, hazy patch with binoculars. The Saturn Nebula (NGC 7009) appears only as a greenish point of light through binoculars, but takes on its Saturn-like appearance when seen through a telescope.

*The vast Helix Nebula is 450 light years from Earth.*

🔭 **Globular Cluster M72**
*A faint globular cluster, M72 can only be resolved into stars through a telescope.*

## SAGITTARIUS *The Archer*    Width

When you look at the constellation Sagittarius, you are looking towards the dense center of our own galaxy, the Milky Way, with its wealth of nebulae and star clusters. Some of the most interesting of these are the Lagoon Nebula (M8), visible to the naked eye, the Trifid (M20) and Omega (M17) nebulae, and star clusters M21, M22, and M23, which are all good binocular subjects.

🔭 **Omega Nebula**
*A telescope will show the Omega Nebula's colorful clouds of gas.*

*M22 is a bright globular cluster.*

*The Lagoon Nebula is visible to the naked eye.*

*Beta (β) is a multiple star.*

## CAPRICORNUS *The Sea Goat*

Capricornus is the smallest zodiacal constellation and one of the least conspicuous. The stars represent a fish-tailed goat. The constellation has one bright star, Deneb Algedi, and just one globular cluster, M30. Keen-eyed observers will see that Alpha (α) is a wide double consisting of two yellowish-colored stars.

Width

*Alpha (α) is a double star.*

Deneb Algedi

🔭 **Globular Cluster M30**
*A small telescope can show the bright nucleus of globular cluster M30.*

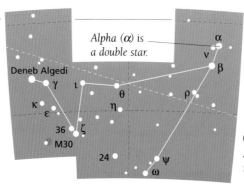

| KEY TO DEEP-SKY OBJECTS | The Milky Way | Galaxy | Globular Cluster | Open Cluster | Diffuse Nebula | Planetary Nebula |
|---|---|---|---|---|---|---|

## SCORPIUS *The Scorpion*

Width 🖐🖐

Like Sagittarius, Scorpius lies in the direction of the Milky Way and, on clear dark nights, rich starfields and dark dust lanes (pp.92–3) can be seen with the naked eye. The brightest star in Scorpius is Antares, a red super-giant that marks the head of the scorpion and is easily spotted. Graffias is also easily seen with the naked eye, but a telescope will show it as two stars.

*These stars mark the sting in the scorpion's tail.*

*Antares is at least 250 times the size of the Sun.*

🔭 **Star Cluster M7**
*This bright, open cluster of about 80 stars covers 1° of sky. It looks like a blur to the naked eye, although with binoculars or a telescope individual stars can be seen.*

## LIBRA *The Scales*

Width 🖐🖐

This constellation occupies a barren area of sky between Scorpius and Virgo; in fact, Libra was once part of Scorpius, and represented the scorpion's claws, an origin still detectable in the Arabic names of its stars; Alpha (α) is called Zubenelgenubi or "southern claw" and Beta (β) is Zubeneschamali or "northern claw." The latter looks quite green, even to the naked eye.

*Delta (δ) is an eclipsing binary (p.87).*

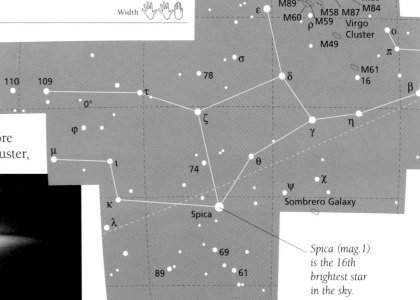

*Binoculars show Zubenelgenubi to be a wide double star.*

👁 **Stars in Libra**
*The stars, Alpha (α), Beta (β), and Sigma (σ) on the right of the picture are the easiest to identify.*

## VIRGO *The Virgin*

Width 🖐🖐🖐

Virgo has more bright galaxies than any other constellation. Many belong to the Virgo Cluster, a group of 2,500 galaxies that extends from Virgo to Coma Berenices (p.121). M84, M86, and M87 can be seen with binoculars, but a telescope is more effective. Brightest of all, but not part of the cluster, is the Sombrero Galaxy (M104), one of the most massive galaxies known.

🔭 **The Sombrero Galaxy**
*A fine sight through a telescope, the Sombrero is a spiral galaxy that lies edge-on to Earth.*

*Spica (mag.1) is the 16th brightest star in the sky.*

| KEY TO STAR MAGNITUDE | ● –1 | ● 0 | ● 1 | ● 2 | ● 3 | • 4 | • 5 | • 6 |

# JANUARY AND FEBRUARY SKY

The night sky in January and February is a superb sight. The most prominent star is Sirius in Canis Major, the brightest star in the entire sky. Gemini and Taurus (pp.106–7) mark the northernmost limit of the path of the zodiac. South of Taurus is magnificent Orion, a part or all of which is visible from any point on Earth. (Key constellations are on pp.112–3.)

## LOCATING THE CONSTELLATIONS

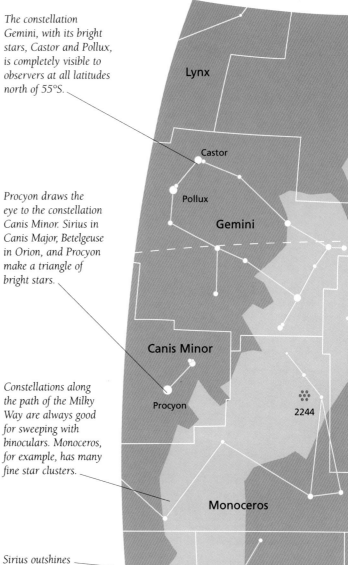

Set the planisphere to find which of these constellations are visible for your location and the time. Northern hemisphere observers will be able to see Auriga overhead and Orion above their horizon. Southern hemisphere observers also get a clear view of Orion, and Canis Major is overhead.

*The constellation Gemini, with its bright stars, Castor and Pollux, is completely visible to observers at all latitudes north of 55°S.*

*Procyon draws the eye to the constellation Canis Minor. Sirius in Canis Major, Betelgeuse in Orion, and Procyon make a triangle of bright stars.*

*Constellations along the path of the Milky Way are always good for sweeping with binoculars. Monoceros, for example, has many fine star clusters.*

*Sirius outshines all other stars in the sky. Its name comes from the Greek for "scorching."*

Lynx · Castor · Pollux · Gemini · Canis Minor · Procyon · 2244 · Monoceros · Sirius · M41 · Canis Major · Puppis

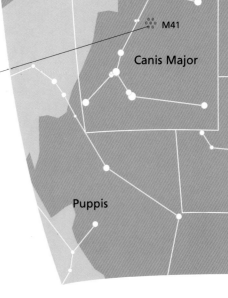

⊙ **January and February sky**
*At the beginning of the year, there are many bright stars in the sky, like those in Orion, on the right of this photograph. Once your eyes have become adapted to the dark, many more stars will be apparent.*

⊙ **M41 in Canis Major**
*Just south of Sirius is star cluster M41. It is a young cluster, only about 100 million years old. Many of its 80 stars are visible through binoculars.*

KEY TO DEEP-SKY OBJECTS |  The Milky Way |  Galaxy |  Globular Cluster |  Open Cluster |  Diffuse Nebula | ◎ Planetary Nebula

Auriga

Perseus

M38

M36

M37

Taurus

*Star cluster M38 has a loose cross shape, when viewed through a telescope.*

### ☀ Star cluster M36 in Auriga
*In the constellation Auriga are three star clusters that lie along the Milky Way. Star cluster M36 is a good subject for viewing with a telescope, through which the individual stars in the cluster can be seen. The brightest of these stars is mag.8, the rest are much fainter, reaching only about mag.13.*

Hyades

Orion

Betelgeuse

*The Hyades star cluster, one of the best in the northern winter sky, contains hundreds of stars. Over 100 are brighter than mag.9.*

*Orion, the Hunter, straddles the celestial equator. His upper body is in the northern celestial sky, and his belt and lower body are in the southern celestial sky.*

M42

Eridanus

### ☾ The Milky Way
*The star-studded Milky Way flows across the sky of the northern hemisphere in winter. It is best seen on a clear, dark night away from city lights. Hundreds of stars can be seen through binoculars.*

Lepus

Columba

Caelum

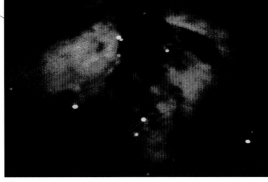

### ☀ The Flame Nebula in Orion
*The constellation Orion has many fine sights. Close to the Orion Nebula is the Flame Nebula. The gas and dust of this nebula makes a misty backdrop for several bright stars. The full beauty of this nebulous patch, however, can be seen only through a telescope.*

### TIPS FOR OBSERVERS

**The Orion Nebula** can be seen with the naked eye. If you have trouble seeing it, use the averted vision technique (p.15).

**Do not look** for faint objects when the Moon is full or close by, because its light is too bright. Wait until the sky is as dark as possible.

**The magnitude figure** given for a star cluster on the following pages is for the combined stellar brightness of the cluster.

| KEY TO STAR MAGNITUDE | | –1 | | 0 | | 1 | | 2 | | 3 | | 4 | | 5 |

# JANUARY AND FEBRUARY: Key Constellations

The key constellations in the months of January and February have an abundance of bright stars and deep-sky objects. The unmistakable shape of Orion dominates the night sky; this constellation is one of the most magnificent in the entire sky. Puppis and Monoceros both lie along the path of the Milky Way, where there are rich starfields and many objects of interest for binocular and telescope observers alike; a number of these also make excellent subjects for astrophotography.

## AURIGA *The Charioteer*                Width

Capella is the most brilliant star in the constellation Auriga and, at mag.0.1, one of the brightest stars in the northern hemisphere sky and one of the easiest to spot. Where the path of the Milky Way runs through Auriga there are three fine open star clusters. They are on the limit of naked-eye visibility but can easily be seen through binoculars as starry patches of light.

### Star clusters M36, M37, and M38
*Seen from left to right, M37, the largest of the clusters, is a tight collection of about 150 stars; star cluster M36 has about 60 stars, and M38 has about 100.*

Capella is a yellow giant star.

M37 is the finest of the trio to observe.

This star is shared with Taurus (p.107).

## ORION *The Hunter*                Width

Orion is a spectacular constellation with many bright stars and deep-sky objects. In the hunter's shoulder, Betelgeuse is a red supergiant (mag.0.5). Rigel, in the left foot, is a blue-white star of mag.0.1. The rest of the figure is formed by Bellatrix and Saiph, with three stars marking his belt, and the Orion Nebula his sword. To the naked eye, this huge cloud of gas and dust looks like a faint patch, but a telescope shows four stars at the nebula's heart.

### The Horsehead Nebula
*The horsehead shape of this nebula is clearly visible only in long-exposure photographs taken through a telescope.*

Three aligning stars form Orion's belt.

The Orion Nebula forms the hunter's sword.

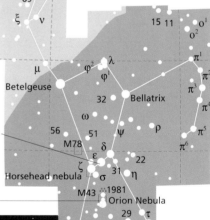

### Stars in Orion
*The bright stars outlining Orion's figure make this one of the most conspicuous constellations in the sky.*

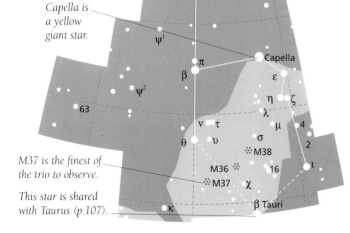

| KEY TO DEEP-SKY OBJECTS | The Milky Way | Galaxy | Globular Cluster | Open Cluster | Diffuse Nebula | Planetary Nebula |
| --- | --- | --- | --- | --- | --- | --- |

## CANIS MAJOR *The Great Dog*

Width 🖐

Canis Major, close to Orion in the sky, represents one of Orion's hunting dogs. The brilliant star Sirius (mag.−1.44) dominates this constellation and the surrounding sky. It is the brightest star in the night sky, but is only about 25 times as powerful as the Sun. It shines so brightly because it is relatively close to Earth. There are other bright stars, notably Adhara, a blue giant (mag.1.5), and Wesen, a yellow supergiant (mag.1.9). Below Sirius, the beautiful open star cluster, M41, is just visible to the naked eye.

*Sirius has a companion star, known as "the Pup," which can be seen with a powerful telescope.*

*Wesen is a yellow supergiant.*

### 👁 Stars in Canis Major
*This constellation has several bright stars, including Sirius, which marks the dog's nose, and Adhara, one of the dog's hindlimbs.*

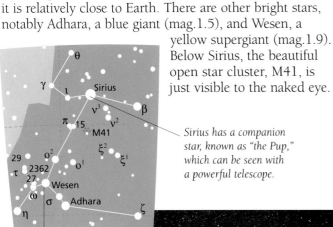

## PUPPIS *The Stern*

Width 🖐

Although this constellation has no bright stars for naked-eye observers, it is crossed by the Milky Way and has a rich array of star clusters that can be seen with binoculars. A good starting point is M47, a scattered open star cluster with about 30 stars. Star cluster, NGC 2451, is also visible through binoculars. The brightest star in this cluster is an orange giant of mag.3.6.

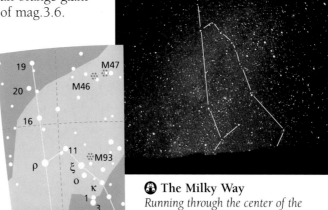

### 🔭 The Milky Way
*Running through the center of the constellation Puppis is the band of the Milky Way with its spectacular starfields and star clusters.*

*NGC 2451 covers an area larger than the Full Moon.*

*L is a wide pair of stars, each visible to the naked eye.*

*V is an eclipsing binary (p.87).*

## MONOCEROS *The Unicorn*

Width 🖐🖐

As Monoceros has so few bright stars of its own, it is best located through finding three brilliant neighboring stars: Procyon in Canis Minor, Betelgeuse in Orion, and Sirius in Canis Major. Lying in the Milky Way, this constellation has many deep-sky objects including the Rosette Nebula, a cluster of about 30 stars visible with binoculars. The surrounding nebula can be seen clearly with a large telescope.

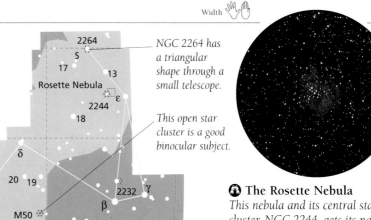

*NGC 2264 has a triangular shape through a small telescope.*

*This open star cluster is a good binocular subject.*

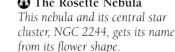

### 🔭 The Rosette Nebula
*This nebula and its central star cluster, NGC 2244, gets its name from its flower shape.*

| KEY TO STAR MAGNITUDE | ⬤ −1 | ⬤ 0 | ⬤ 1 | ⬤ 2 | • 3 | • 4 | • 5 | · 6 |

113

Left margin (vertical): JANUARY FEBRUARY **MARCH** APRIL MAY JUNE JULY AUGUST SEPTEMBER OCTOBER NOVEMBER DECEMBER

# MARCH AND APRIL SKY

Leo (p.106) is prominent at this time of year. For northern observers, this constellation signals the arrival of spring, and for those in the southern hemisphere, fall. Leo's bright star, Regulus, is easily seen with the naked eye. Southern observers will be able to observe the spectacular constellation Vela, lying in the Milky Way, and the whole of Hydra. (Key constellations are on pp.116–7.)

## LOCATING THE CONSTELLATIONS

Set the planisphere to find which of these constellations are visible for your location and the time. Northern hemisphere observers see the constellations Leo and Cancer high in the sky. Southern hemisphere observers see the constellation Vela and the rich path of the Milky Way overhead.

*This pattern of stars in Leo resembles a back-to-front question mark or farmer's sickle.*

*The bright star, Regulus, in Leo lies almost on the path of the ecliptic.*

*The triangle of galaxies M95, M96, and M105 can be seen through a small telescope.*

Constellation map labels: Ursa Major, Leo Minor, Leo, Regulus, M105, M96, M95, Virgo, Sextans, Crater, Ghost of Jupiter, Hydra, Antlia, Eight-burst Nebula

⊗ **The Ghost of Jupiter Nebula in Hydra**
*Through a large telescope, this planetary nebula, NGC 3242, looks like a human eye, but its name comes from its planet-like disk appearance when it is seen through a small telescope.*

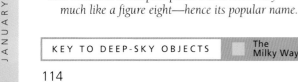

⊗ **The Eight-burst Nebula in Vela**
*This planetary nebula, NGC 3132, can be seen through a small telescope as a rounded disk. Through a large telescope it seems to be arranged in a number of rings intertwined in a complex pattern that looks very much like a figure eight—hence its popular name.*

| KEY TO DEEP-SKY OBJECTS | The Milky Way | Galaxy | Globular Cluster | Open Cluster | Diffuse Nebula | Planetary Nebula |
|---|---|---|---|---|---|---|

Lynx

2683

Cancer

M44

M48

Hydra

Pyxis

Puppis

Vela

### ⚙ NGC 2683 in Lynx

*The barred spiral galaxy, NGC 2683, is a spectacular sight through a telescope. It lies almost edge-on to Earth which unfortunately makes its beautiful spiral structure difficult to see. The bright central area is the galaxy's nucleus which is densely packed with stars.*

### 👁 Leo and Leo Minor

*The bright stars forming the outline of the lion's body make this an easy constellation to pick out in the sky. At center bottom is the bright star, Regulus, which marks the lion's chest. Above Leo are the fainter stars of Leo Minor.*

*The path of the Sun cuts through the zodiacal constellation Cancer.*

### 🔭 M48 in Hydra

*Star cluster M48 lies in a barren area of the constellation Hydra, some distance away from the snake's body. It contains about 80 stars, the brightest of which are mag.9. Many of these stars can be seen as individual stars through binoculars or a small telescope.*

### TIPS FOR OBSERVERS

**Binoculars** have a wider field of view than telescopes, so the complete spread of a star cluster, such as M48 in Hydra, can be seen with binoculars, compared to only part with a telescope. **Although** a constellation is above the horizon, some of its stars may not be visible at a particular latitude. To see them, they should be high in the sky where the contrast between starlight and dark sky is greatest.

JANUARY FEBRUARY **MARCH** APRIL MAY JUNE JULY AUGUST SEPTEMBER OCTOBER NOVEMBER DECEMBER

KEY TO STAR MAGNITUDE   ● −1   ● 0   ● 1   ● 2   ● 3   • 4   · 5   · 6

# MARCH AND APRIL: Key Constellations

The key constellations in the March and April sky are extremely varied. Vela is a rich constellation, particularly where it is crossed by the Milky Way. Hydra is extensive, and has some good sights along its entire length for observers with binoculars and small telescopes. Leo Minor is a small companion of the zodiacal constellation Leo, but it is rather insignificant compared with Leo. Sextans and Lynx are both faint constellations requiring good observation conditions for naked-eye viewing.

## VELA *The Sail*
Width

The bright path of the Milky Way runs through Vela and, with it, an array of open star clusters, which make good binocular subjects and the so-called "Eight-Burst Nebula" (NGC 3132), one of the brightest planetary nebulae in the sky. This nebula can be spotted with the naked eye as a hazy round patch of light, but its disk shape can only be seen clearly through a telescope or as a photographic image. Photographs reveal the nebula's intricate loops.

This globular cluster looks like a disk of stars through binoculars.

Open cluster NGC 2547 is visible with binoculars.

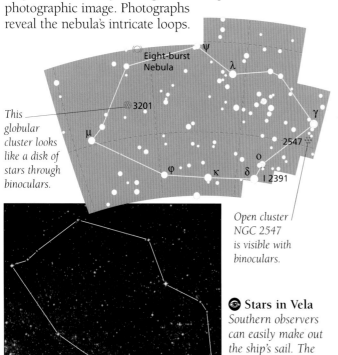

● **Stars in Vela**
Southern observers can easily make out the ship's sail. The brightest stars stand out against the path of the Milky Way.

## LEO MINOR *The Little Lion*
Width

The constellation Leo Minor is much smaller and far less interesting than its companion, Leo. Leo Minor has no bright stars and, although there are galaxies that lie within its boundaries, they are beyond the limits of most amateur observing equipment. The variable star R is an interesting object for binocular observers. It can be viewed over 372 days as its brightness varies from mag.6.3 to mag.13.2.

*Beta (β) is a double star but looks single to the naked eye.*

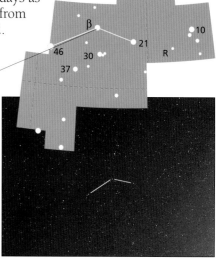

● **Stars in Leo Minor**
*Representing a lion cub, the constellation Leo Minor is quite faint and therefore difficult to pick out.*

## SEXTANS *The Sextant*
Width

This constellation lies on the celestial equator (pp.16–17), just to the south of Leo. It can be difficult to make out in the sky because it has no really bright stars; only five in the constellation are brighter than mag.5.5. The brightest star is Alpha (α), which is mag.4.5.

*This wide pair of double stars is visible with binoculars.*

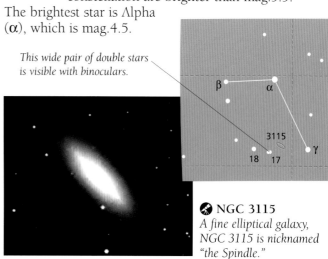

⊘ **NGC 3115**
*A fine elliptical galaxy, NGC 3115 is nicknamed "the Spindle."*

| KEY TO DEEP-SKY OBJECTS | The Milky Way | Galaxy | Globular Cluster | Open Cluster | Diffuse Nebula | Planetary Nebula |

## HYDRA *The Water Snake*

Width

This is the largest constellation in the night sky but it is not easy to see in its entirety, and is best viewed piecemeal. The snake's head is formed by a tight group of six stars of mag. 3 and mag.4. The first part of its body is denoted by the bright star Alphard (mag.2), the brightest star in the constellation, then a "w" of stars make up the next portion of the snake's body, which then slithers under the constellation Sextans and finally straightens into the tail under Virgo and Libra (p.109). Features of interest include binocular cluster M48, on the very edge of the constellation's boundary.

*This planetary nebula is visible with a small telescope.*

*Alphard (mag.2) is an orange giant star.*

*R is a variable giant star of the Mira type (p.87).*

*M83, a spiral galaxy, lies face-on to Earth.*

*M68 is a globular cluster that can be resolved into stars with a telescope.*

👁 **Stars in Hydra**
*The stars in the head of the snake are the brightest, and are visible to the naked eye. This part of Hydra shares a border with the zodiacal constellation Cancer.*

✦ **Spiral Galaxy M83**
*This galaxy's shape is visible with a large telescope. Small telescopes and binoculars show it as a light patch with a bright center.*

## LYNX *The Lynx*

Width

At mag.4, even the brightest stars in Lynx are faint, and the constellation occupies a barren patch of sky. There are deep-sky objects within the constellation's boundaries but only one, NGC 2683, is easily visible with amateur equipment. This is a barred-spiral galaxy that is positioned edge-on to Earth, making its spiral structure very difficult to view clearly, even with quite large telescopes.

*This galaxy appears as an oblong patch of light through binoculars.*

👁 **Stars in Lynx**
*The stars in this constellation are very indistinct. Its brightest star, alpha (α) is the most easily seen with the naked eye.*

| KEY TO STAR MAGNITUDE | –1 | 0 | 1 | 2 | 3 | 4 | 5 | 6 |
|---|---|---|---|---|---|---|---|---|

117

# MAY AND JUNE SKY

The zodiacal constellations, Virgo and Libra (p.109), are prominent in the sky at this time of year. Virgo is a large constellation that lies just north of the ecliptic. Straddling Virgo, and its neighbor Coma Berenices, is the Virgo–Coma Cluster, an immense group of galaxies, many of which can be seen from Earth. The brightest stars in the sky at this time are Arcturus in Boötes and Spica in Virgo. (Key constellations are on pp.120–1.)

*Despite its common name, "Northern Crown," Corona Borealis can be seen from all latitudes north of 50°S.*

*This is the top half of Serpens, which is the only star pattern to be cut in half by another constellation, Ophiuchus.*

*Virgo is only entirely visible from latitudes between 68°N and 76°S.*

## LOCATING THE CONSTELLATIONS

Set the planisphere to find which of these constellations are visible for your location and the time. Observers in the northern hemisphere will be able to see the constellations Boötes and Corona Borealis overhead, and southern hemisphere observers will be well placed to view Corvus and Lupus.

⊙ **The Aurora Borealis seen at 2:50 A.M., May 1, 1990 in Scotland**
*Observers at latitudes north of about 55°N can see the Aurora Borealis (pp.72–3) several times a year. Displays are not easy to predict but, as they can take place over two to three nights, you may see one. This display was particularly good and lasted for several hours.*

Boötes

Corona Borealis

Serpens Caput

Libra

Lupus

| KEY TO DEEP-SKY OBJECTS | The Milky Way | Galaxy | Globular Cluster | Open Cluster | Diffuse Nebula | Planetary Nebula |

118

Whirlpool
Galaxy

Canes Venatici

M3

Coma
Berenices

M100

Virgo

Corvus

Hydra

Centaurus

5139

### The Whirlpool Galaxy in Canes Venatici

*The Whirlpool Galaxy (M51) lies face on to our line of sight on Earth. Most binoculars will show it as a circular patch of light, but its whirlpool shape becomes obvious when viewed through a large telescope. An irregular-shaped companion, NGC 5195, is connected to the main galaxy by an extended spiral arm.*

### M100 in Coma Berenices

*A large telescope is needed to see the spiral arms of galaxy M100. With a smaller telescope this member of the Virgo Cluster (p.109) looks like a patch of light with a bright center, and can easily be mistaken for a globular cluster.*

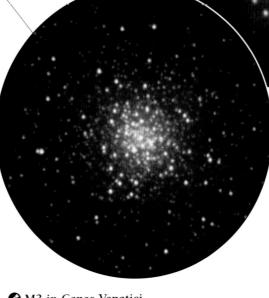

### TIPS FOR OBSERVERS

**When observing** the Virgo Cluster, make sure you know which galaxy you are observing before moving on to the next one. It is easy to get confused. **The scattered** star cluster that makes up the hair in Coma Berenices (Berenice's hair) covers about 5° of sky. Low-power and wide-field instruments must be used to look at the group or the clustering effect will be lost.

### M3 in Canes Venatici

*This is one of the brightest star clusters in the sky, containing about half-a-million stars. It can be seen as a glowing star-like cloud through binoculars. Some of its stars can be resolved through telescopes with 4-in (100-mm) apertures and above.*

| KEY TO STAR MAGNITUDE | −1 | 0 | 1 | 2 | 3 | 4 | 5 |

# MAY AND JUNE: Key Constellations

Two of these constellations, Canes Venatici and Coma Berenices, contain numerous galaxies, many of which can be seen with binoculars. Canes Venatici has the lovely Whirlpool Galaxy, which is one of the most beautiful in the sky. Because it lies face-on to Earth, its attractive spiral structure can clearly be seen. In Coma Berenices is the distinctive Black Eye Galaxy.

## LUPUS *The Wolf* Width

Lupus lies between the constellations Scorpius (p.109) and Centaurus (p.102). Once these constellations have been located, it is then easy to distinguish the stars making the wolf's shape. Where the path of the Milky Way crosses Lupus, there are several star clusters. Open cluster NGC 5822, with over 100 stars, is probably the most attractive of them.

*Kappa (κ) is a double star; the two can easily be seen with a small telescope.*

**Star Cluster 5822**
*This open star cluster can be seen with the naked eye, and it is an ideal target for observers with binoculars or small telescopes.*

## CORONA BOREALIS *The Northern Crown* Width

Corona Borealis is tiny, but it is an easy constellation to see because of its distinctive tiara-like shape. The brilliant star Gemma marks the crown's central jewel at mag.2.2, and within the crescent shape is the variable star R that fluctuates between mag.6 and mag.14.

## BOÖTES *The Herdsman* Width

The brilliant star Arcturus (mag.0) leads the eye to this constellation; it is an orange-red star and the fourth brightest in the sky. The name Arcturus comes from the Greek for "bearkeeper;" and Boötes is the herdsman driving the constellation Ursa Major (the Great Bear) across the sky. The Quadrantids meteor shower (pp.74–5) radiates from a certain area of the Boötes constellation in January every year.

*The meteor shower, named the Quadrantids after a now defunct constellation, radiates from this part of the sky.*

*A telescope will show that Izar (ε) is a close pair of mag.3 and mag.5 stars.*

*Arcturus is the brightest star north of the celestial equator and has a warm tint when seen with the naked eye.*

**Stars in Corona Borealis and Boötes**
*The kite-shaped outline of Boötes can be picked out in the center of the picture; Corona Borealis is the group of stars forming a semicircle to its left.*

| KEY TO DEEP-SKY OBJECTS | | The Milky Way | | Galaxy | | Globular Cluster | | Open Cluster | | Diffuse Nebula | | Planetary Nebula |

## CORVUS *The Crow*

Width 👆

This is a small constellation in an empty part of the sky that is just south of Virgo (p.109). It represents a crow holding on to the body of Hydra (p.117). At the edge of the constellation is a pair of colliding galaxies known as the Antennae. Through a small telescope they look like a doughnut with a bite taken out.

👁 **Stars in Corvus**
*The bright stars of Corvus can be seen with the naked eye. They make a trapezoid shape.*

## CANES VENATICI *The Hunting Dogs*

Width ✋✋

A lot of imagination is needed to trace out Boötes' two hunting dogs. Only two stars, Cor Caroli and Beta (β) are immediately noticeable and are best located by looking under Ursa Major's tail (p.98). Cor Caroli, the brighter star of the two, is mag.2.9 and has a mag.5 companion. Two spiral galaxies, the Whirlpool and M94 look like round blurs of light in a small telescope. Globular cluster, M3, is on the limit of naked-eye visibility.

*Cor Caroli is a double star easily resolved with a small telescope.*

👁 **Stars in Canes Venatici**
*The stars Alpha (α) and Beta (β) are visible in this naked-eye view. Alpha is called Cor Caroli or Charles' heart, named in 1673 after Charles I of England.*

## SERPENS CAPUT *The Serpent's Head*

Width 👆

Serpens Caput is not a constellation in its own right but one half of Serpens, the Serpent, which is cut in two by Ophiuchus (p.105). Serpens Cauda, the tail of the Serpent, is seen later in the year (p.125). A triangle of stars marks the head and a crooked line its upper body. In an otherwise barren area of sky, M5 is a spectacular star cluster.

*Delta (δ) is a binary (p.86), the two parts of which can be resolved with a small telescope.*

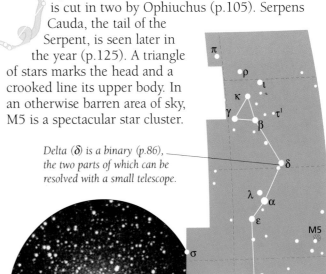

🔭 **Globular Cluster M5**
*Some of the half a million stars in this cluster can be resolved through a telescope. The rest are tightly packed in the center of the globular cluster.*

## COMA BERENICES *Berenice's Hair*

Width ✋

A faint constellation with only a few noteworthy stars, Coma Berenices shares a border with Virgo (p.109). Many of the galaxies belonging to the Virgo–Coma Cluster (sometimes also called the Virgo Cluster) straddle these two constellations. The most impressive galaxy is NGC 4565.

🔭 **The Black Eye Galaxy**
*The dark cloud of dust across the bright center of spiral galaxy M64 can be seen only with a telescope.*

| KEY TO STAR MAGNITUDE | | –1 | | 0 | | 1 | | 2 | | 3 | | 4 | | 5 | | 6 |

# JULY AND AUGUST SKY

The magnificent zodiacal constellations Sagittarius and Scorpius (pp.108–9) are prominent in the July and August night sky. They lie in a rich part of the Milky Way and contain many stellar objects for all observers, especially for those in the southern hemisphere. Some northern hemisphere observers will not see Sagittarius and Scorpius, but will be in a good position to view Hercules, Aquila, and Lyra. (Key constellations are on pp.124–5.)

## LOCATING THE CONSTELLATIONS

Set the planisphere to find which of these constellations are visible for your location and time. For observers in the northern hemisphere, Hercules and Hydra are overhead. Southern hemisphere observers see the zodiacal constellations Scorpius and Sagittarius, and the rich path of the Milky Way.

*This planetary nebula in Lyra, called the Ring Nebula, is a fine sight through binoculars.*

*Lyra is one of the most conspicuous constellations of the northern hemisphere sky and is completely visible to observers located north of 42°S.*

**M56 in Lyra**
*On the edge of the constellation Lyra, is the globular cluster M56, which looks like a fuzzy star with a bright center. The stars on the outside of the cluster can be resolved with a 6-in (150-mm) telescope.*

**The Wild Duck Cluster in Scutum**
*This star cluster, M11 in Scutum, is named the Wild Duck because some observers think it resembles a flight of ducks. It is made up of hundreds of stars, many of which can be seen through a small telescope.*

Cygnus
Lyra
Ring Nebula
M56
M27
Vulpecula
Sagitta
Serpens Cauda
Aquila
Wild Duck Cluster
Eagle Nebula
Scutum
M16
M17
M22
M8
Sagittarius
Corona Australis
Telescopium

| KEY TO DEEP-SKY OBJECTS | The Milky Way | Galaxy | Globular Cluster | Open Cluster | Diffuse Nebula | Planetary Nebula |
|---|---|---|---|---|---|---|
| |  |  |  |  |  |  |

Hercules

Corona Borealis

Ophiuchus

Serpens Caput

M13

M12

M10

Scorpius

M6

M7

Norma

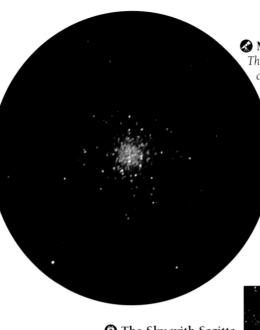

### 🛰 M13 in Hercules
*This is one of the finest globular star clusters in the sky. It ranks only behind Omega Centauri (p.101) and 47 Tucanae (p.103). Through binoculars it looks very much like a fuzzy cloud of about half the apparent width of the Full Moon, but through a 4-in (100-mm) telescope or larger many of the cluster's hundreds of thousands of stars become clearly visible.*

### 🌐 The Sky with Sagitta
*The arrow shape of the constellation Sagitta is often obvious through binoculars, but sometimes, as in this photograph, it is difficult to see against the star-studded background of the Milky Way.*

*Two globular clusters, M10 and M12 in Ophiuchus are visible simultaneously; M12 is the more open of the two.*

### 🌐 M16 in Serpens and M17 in Sagittarius
*The bright patches in this photograph fall in two constellations. At the top is star cluster, M16, which is surrounded by the Eagle Nebula in Serpens Cauda; and below is M17, the Omega Nebula in Sagittarius.*

## TIPS FOR OBSERVERS

**Do not mistake** the Wild Duck Cluster (M11) for a globular cluster. The stars of M11 are densely packed and have the appearance through binoculars of a globular cluster whose stars have not been resolved.

**A bright bar** of cloud is seen first when observing the Omega Nebula, M17. To see the nebula's fainter part, which curves away from the bar, use averted vision (p.15).

| KEY TO STAR MAGNITUDE | ● −1 | ● 0 | ● 1 | ● 2 | ● 3 | · 4 | · 5 |
|---|---|---|---|---|---|---|---|

# JULY AND AUGUST: Key Constellations

The key constellations in the July and August night sky range in size from the very large, such as Hercules, depicting the hero of Greek myth, to the very small, such as Sagitta. All these constellations contain interesting objects for observers with any type of instrument. Hercules has M13, the brightest globular cluster in the northern hemisphere sky, and tiny Sagitta has an interesting quick-changing eclipsing binary star (p.87). The two brilliant stars, Vega and Altair in the constellations Lyra and Aquila respectively, outshine all other stars in the sky at this time of year. With Deneb in Cygnus (p.129), Vega and Altair form a distinctive triangle of bright stars.

## SAGITTA *The Arrow*

Width

Sagitta is one of the smallest constellations in the sky. It lies in the Milky Way with the constellations Vulpecula and Cygnus to the north, and Aquila to the south. It has no bright stars but at its heart is loose globular star cluster M71, a rewarding binocular or telescope subject, situated halfway between Gamma (γ) and Delta (δ).

γ    δ  α
M71    β

**⊛ Star Cluster M71**
*Although M71 looks like an open cluster, it is actually a globular cluster. A telescope can resolve its stars.*

## HERCULES *Hercules*

Width

Hercules is the fifth-largest constellation in the sky. The head of the giant points towards the south celestial pole and his feet to the north celestial pole. The box shape of the giant's body is known as the "Keystone of Hercules" and is made up of four bright stars. Beta (β), a yellow giant of mag.2.8, marks the giant's knee; it is the brightest star in the constellation. Rasalgethi, in the head, is a variable star (p.87) that alters in brightness from mag.3 to mag.4 every 90 days.

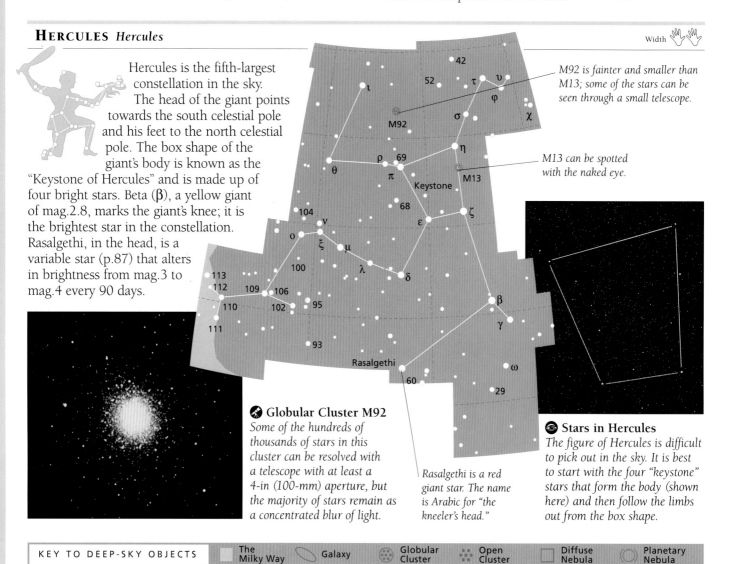

M92 is fainter and smaller than M13; some of the stars can be seen through a small telescope.

M13 can be spotted with the naked eye.

42
52    τ    υ
φ
σ              χ
M92
η
ρ    69
θ         M13
π    Keystone
68
104         ζ
ν         ε
o    ξ
μ    λ    δ
100
113
112  109  106
110    102    95    β
111                γ
93
Rasalgethi    ω
60    29

**⊛ Globular Cluster M92**
*Some of the hundreds of thousands of stars in this cluster can be resolved with a telescope with at least a 4-in (100-mm) aperture, but the majority of stars remain as a concentrated blur of light.*

*Rasalgethi is a red giant star. The name is Arabic for "the kneeler's head."*

**◉ Stars in Hercules**
*The figure of Hercules is difficult to pick out in the sky. It is best to start with the four "keystone" stars that form the body (shown here) and then follow the limbs out from the box shape.*

| KEY TO DEEP-SKY OBJECTS | The Milky Way | Galaxy | Globular Cluster | Open Cluster | Diffuse Nebula | Planetary Nebula |
|---|---|---|---|---|---|---|

## LYRA *The Lyre*

Width

The brilliant star Vega makes Lyra an easy constellation to find. Vega is a blue-white star of mag.0 and the fifth brightest star in the sky. The other stars are fainter but are still interesting. Sheliak is a variable star (p.87), that alters in brightness from mag.3.4 to mag.4.3. Epsilon (ε) is a quadruple star and the Ring Nebula (M57) is notable for its smoke-ring appearance.

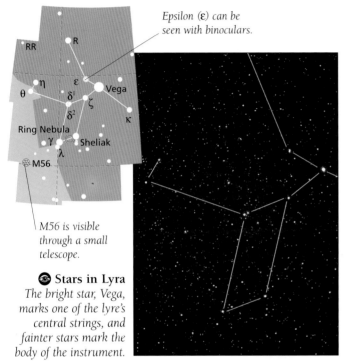

*Epsilon (ε) can be seen with binoculars.*

*M56 is visible through a small telescope.*

### ◉ Stars in Lyra
*The bright star, Vega, marks one of the lyre's central strings, and fainter stars mark the body of the instrument.*

## SERPENS CAUDA *The Serpent's Tail*

Width

The constellation Serpens, the serpent, is divided in two by Ophiuchus (p.105). The head of Serpens is discussed on p.121. Serpens Cauda, the tail, is not so distinctive, but it is crossed by the Milky Way and is home to star cluster M16 and its surrounding nebula (pp.88–9), known as "the Eagle" due to the distinctive shape of its central dust lane. Binoculars with filters will reveal the gas clouds.

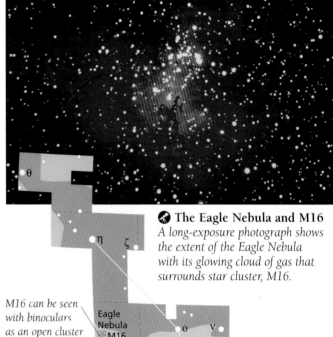

### ◉ The Eagle Nebula and M16
*A long-exposure photograph shows the extent of the Eagle Nebula with its glowing cloud of gas that surrounds star cluster, M16.*

*M16 can be seen with binoculars as an open cluster of about 60 stars.*

## AQUILA *The Eagle*

Width

In Greek mythology, Aquila was the eagle that transported the thunderbolts of the god Zeus. The name of the constellation's brightest star, Altair, comes from the Arabic for "flying eagle." This star completes the triangle of bright stars made by Deneb in Cygnus and Vega in Lyra. Part of Aquila lies along the path of the Milky Way where there are dense starfields, particularly where it borders Scutum.

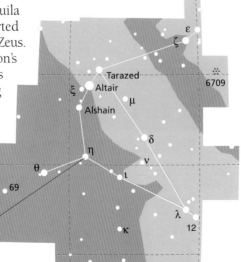

*Eta (η) is a Cepheid variable that can be seen with binoculars.*

### ◉ Stars in Aquila
*The bright star, Altair, is easily located at the center top of the shape in the photograph. To either side of Altair are the bright stars, Alshain and Tarazed.*

KEY TO STAR MAGNITUDE    ⬤ -1   ⬤ 0   ● 1   ● 2   ● 3   • 4   • 5   • 6

*(left margin vertical text)* JANUARY FEBRUARY MARCH APRIL MAY JUNE JULY AUGUST **SEPTEMBER OCTOBER** NOVEMBER DECEMBER

# SEPTEMBER AND OCTOBER SKY

Capricornus and Aquarius (p.108) are prominent in the night sky at this time of year. Aquarius is the most northerly of the two constellatons, but parts or all of it can be seen throughout the world. Capricornus is only visible to those south of 62°N. Pegasus heralds the approach of fall for northern hemisphere observers and Spring for those in the southern hemisphere.

## LOCATING THE CONSTELLATIONS

Set the planisphere to find which of these constellations are visible for your location and the time. Northern hemiphere observers will see Cygnus and the path of the Milky Way overhead. Southern hemisphere observers will see Pegasus above their northern horizon.

🪐 **Stephan's Quintet in Pegasus**
*This quintet of galaxies can only be viewed through a telescope. Taken together, the five have an average magnitude of 13, and can be seen as blurs of light. Their various galaxy shapes can only be made out in larger telescopes.*

 **Piscis Austrinus and Aquarius in the night sky**
*This picture shows Piscis Austrinus (bottom right) with its bright star, Fomalhaut. Aquarius is above it, and to the left of Aquarius is Pisces with part of Cetus below.*

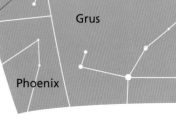

*The constellation Lacerta only rises above the horizon for observers living north of 33°S.*

*The Great Square of Pegasus is an easy shape to recognize. It is high in the sky for northern observers.*

*Fomalhaut in Piscis Austrinus is the 18th brightest star the night sky.*

Andromeda
Lacerta
Stephan's Quintet
Pegasus
Pisces
Aquarius
Sculptor
Formalhaut
Piscis Austrinus
Grus
Phoenix

| KEY TO DEEP-SKY OBJECTS |  The Milky Way | Galaxy | Globular Cluster | Open Cluster | Diffuse Nebula | Planetary Nebula |

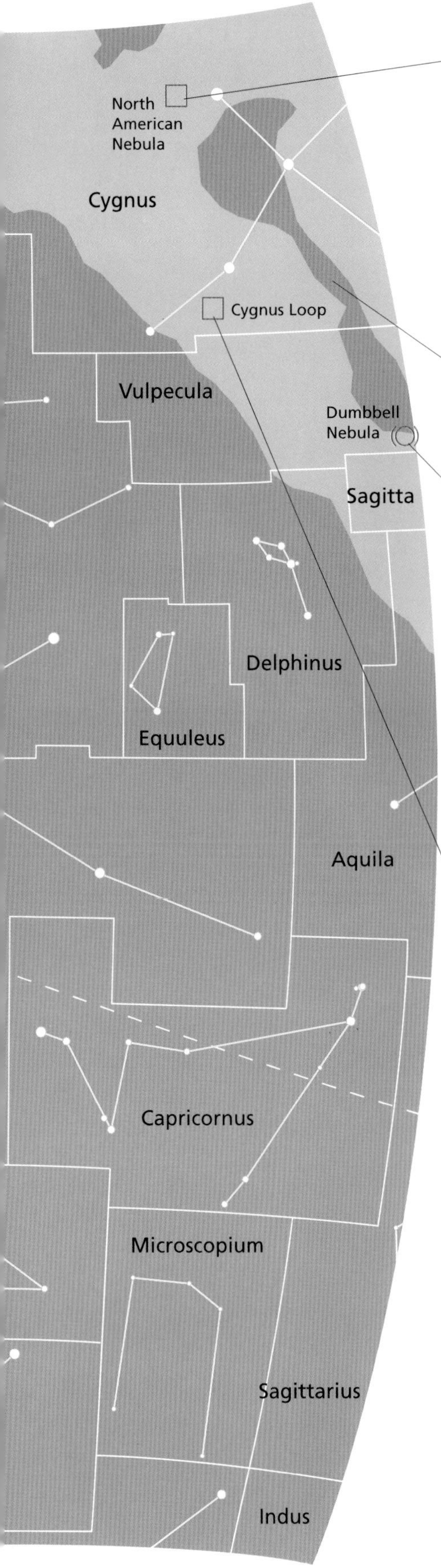

North
American
Nebula

Cygnus

Cygnus Loop

Vulpecula

Dumbbell
Nebula

Sagitta

Delphinus

Equuleus

Aquila

Capricornus

Microscopium

Sagittarius

Indus

### ☄ The North American Nebula in Cygnus

*This striking nebula (NGC 7000), the shape of which is reminiscent of North America, is seen to best effect in a long-exposure photograph. This nebula is quite difficult to see with the naked eye because of the background brightness of our galaxy, the Milky Way.*

*A dark dust lane known as the Cygnus Rift cuts through the bright path of the Milky Way.*

### ☄ The Dumbbell Nebula in Vulpecula

*Once spotted, the magnificent sight of nebula M27 is difficult to forget. It is one of the finest planetary nebulae in the night sky and can be seen through binoculars, but a telescope is needed to see the dumbbell shape and its color.*

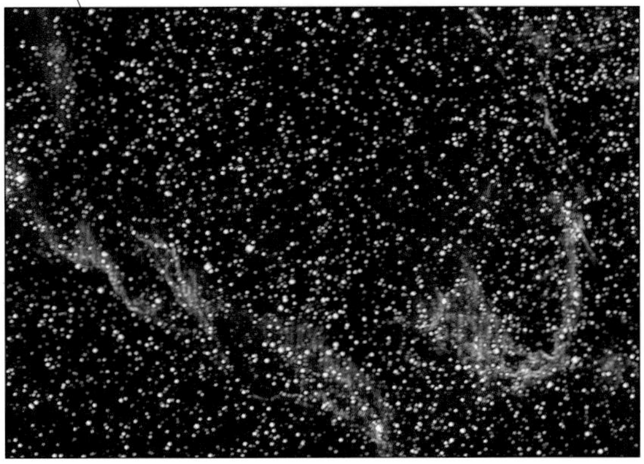

### TIPS FOR OBSERVERS

**A 6-in (150-mm) telescope** is needed to see the hourglass shape of the Dumbbell Nebula, M27, in Vulpecula and its lovely blue-green color.

**The Helix Nebula,** NGC 7293 in Aquarius, appears large in the sky, because it is the nearest planetary nebula to Earth. Because its light is spread out, it is best seen through binoculars and not a telescope, with its narrow field of view.

### ☄ The Cygnus Loop

*The bright streaks of gas and dust that litter this starfield are the remains of a star that blew itself apart at the end of its life, an event known as a supernova (p.89). Some of the brightest streams can be seen with binoculars.*

JANUARY FEBRUARY MARCH APRIL MAY JUNE JULY AUGUST SEPTEMBER OCTOBER NOVEMBER DECEMBER

KEY TO STAR MAGNITUDE   ● −1   ● 0   ● 1   ● 2   ● 3   ● 4   ● 5

# SEPTEMBER AND OCTOBER: Key Constellations

In the final quarter of the year, the Square of Pegasus is prominent in the night sky. Near to its bright star Enif, marking the horse's nose, is the magnificent globular star cluster, M15. The constellation Cygnus becomes rapidly familiar to northern observers because its shape so resembles a swan flying along the Milky Way. Delphinus is also easy to recognize because of its dolphin-like shape. Fomalhaut in Piscis Austrinus is one of the brightest stars in the sky, and Vulpecula, although not a conspicuous constellation, has the lovely Dumbbell nebula.

## PEGASUS *Pegasus*                     Width 🖐🖐

This large constellation represents the head and front part of Pegasus, the winged horse of Greek mythology. Pegasus is easily found by locating the distinctive Great Square, a box of four stars that includes Alpheratz, a star that was once shared by Pegasus and Andromeda but is now officially in Andromeda (p.132). The large spiral galaxy in Pegasus, NGC 7331, can be viewed through binoculars, but the galaxies surrounding Pegasus, including those known as Stephan's Quintet, need a telescope to bring them into view.

## DELPHINUS *The Dolphin*                Width ✋

Although the stars of Delphinus are not bright, the constellation is conspicuous because of its distinctive, dolphin-like shape that stands out in the sky between Aquila and Pegasus. Its two brightest stars, Sualocin and Rotanev, are mag.3.8 and mag.3.6 respectively.

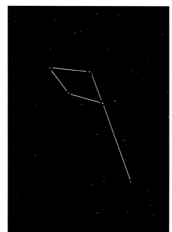

*Gamma (γ) is a double star.*

### 👁 Stars in Delphinus
*The neat shape of Delphinus makes it a pleasing naked-eye subject. The four stars that form the head are known as "Job's Coffin."*

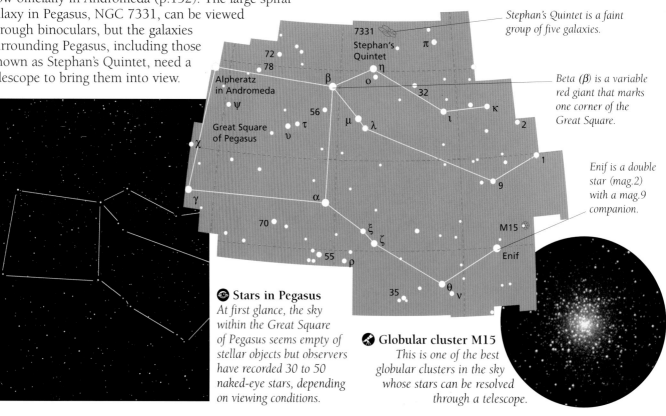

*Stephan's Quintet is a faint group of five galaxies.*

*Beta (β) is a variable red giant that marks one corner of the Great Square.*

*Enif is a double star (mag.2) with a mag.9 companion.*

### 👁 Stars in Pegasus
*At first glance, the sky within the Great Square of Pegasus seems empty of stellar objects but observers have recorded 30 to 50 naked-eye stars, depending on viewing conditions.*

### ⚙ Globular cluster M15
*This is one of the best globular clusters in the sky whose stars can be resolved through a telescope.*

| KEY TO DEEP-SKY OBJECTS | ▦ The Milky Way | ⬭ Galaxy | ⊛ Globular Cluster | ⁘ Open Cluster | ▢ Diffuse Nebula | ◎ Planetary Nebula |

## CYGNUS *The Swan*

Width

Five bright stars forming the shape of a cross make Cygnus an easily recognized constellation. The path of the Milky Way runs through Cygnus, so binocular observers have much to see. NGC 6826 is a planetary nebula, often called the "Blinking Planetary," and NGC 6992 is part of the Veil Nebula (p.88), the remains of a supernova.

*Deneb forms part of a triangle of bright stars.*

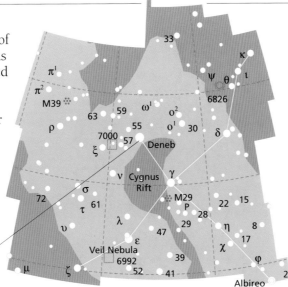

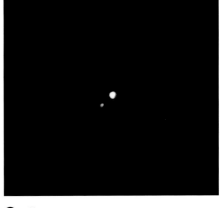

🔭 **Albireo**
*This double star is easily resolved with a telescope, and the colors of the stars, yellow and blue-green, can then be observed.*

## PISCIS AUSTRINUS *The Southern Fish*

Width

The brilliance of the star Fomalhaut (Arabic for "fish's mouth") makes Piscis Austrinus easily spotted. Fomalhaut is among the 20 brightest stars in the night sky, and the remainder of the stars that make up the constellation are all much fainter and undistinguished.

*Beta (β) is a wide double star.*

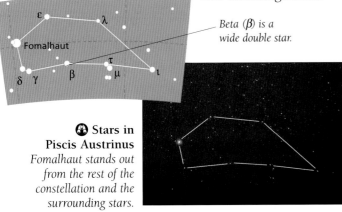

👁🔭 **Stars in Piscis Austrinus**
*Fomalhaut stands out from the rest of the constellation and the surrounding stars.*

## LACERTA *The Lizard*

Width

This constellation is small and insignificant except for two open star clusters, NGC 7243 and NGC 7209, both of which can be seen with binoculars. In the past century, there have been three bright novae (p.85), visible to the naked eye, in Lacerta that have given the constellation further interest.

👁 **Stars in Lacerta**
*The brightest stars in Lacerta are only mag.4, and this makes it difficult to pick out.*

*The stars in this cluster can be seen with a telescope.*

## VULPECULA *The Fox*

Width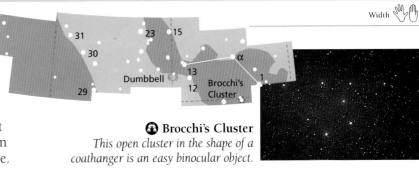

This is not an easy constellaton to find, but it contains a remarkable deep-sky object, a planetary nebula called the Dumbbell (M27). The name comes from the double-lobed shape of the gas cloud that was thrown off by a dying star. The lobes can be seen clearly only through a large telescope.

🔭 **Brocchi's Cluster**
*This open cluster in the shape of a coathanger is an easy binocular object.*

| KEY TO STAR MAGNITUDE | ● −1 | ● 0 | ● 1 | ● 2 | ● 3 | ● 4 | • 5 | • 6 |

Vertical left margin:
NOVEMBER  DECEMBER  OCTOBER  SEPTEMBER  AUGUST  JULY  JUNE  MAY  APRIL  MARCH  FEBRUARY  JANUARY

# NOVEMBER AND DECEMBER SKY

The zodiacal constellations Aries and Pisces (p.107) appear rather faint; however Perseus, Andromeda, and Triangulum have many bright stars and deep-sky objects. Cetus and Eridanus are extensive constellations, and can be seen in their entirety only by southern hemisphere observers. (Key constellations are on pp.132–3.)

## LOCATING THE CONSTELLATIONS

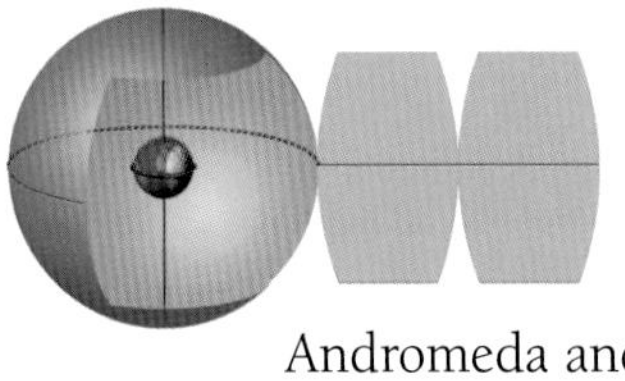

Set the planisphere to find which of these constellations are visible for your location and the time. For northern hemisphere observers, Andromeda and Perseus are high in the sky, and for southern hemisphere observers, Cetus is overhead followed later in the period by Eridanus.

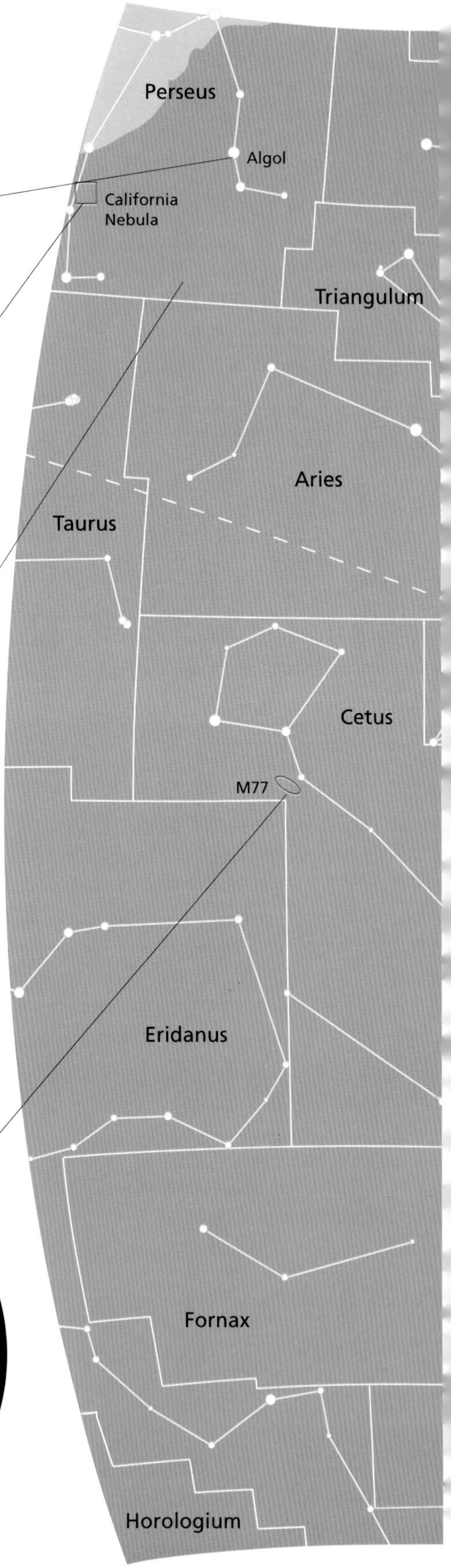

*Algol in Perseus was the first eclipsing binary star (p.87) to be discovered.*

*Observers north of 31°S can see the whole of the constellation Perseus above the horizon.*

### ⓐ The California Nebula in Perseus
*With patience, the observer with binoculars can see nebula NGC 1499 which looks like the State of California. The cloud of dust and gas of the nebula is mag.6, but the light is spread out so the surface brightness is around mag.14.*

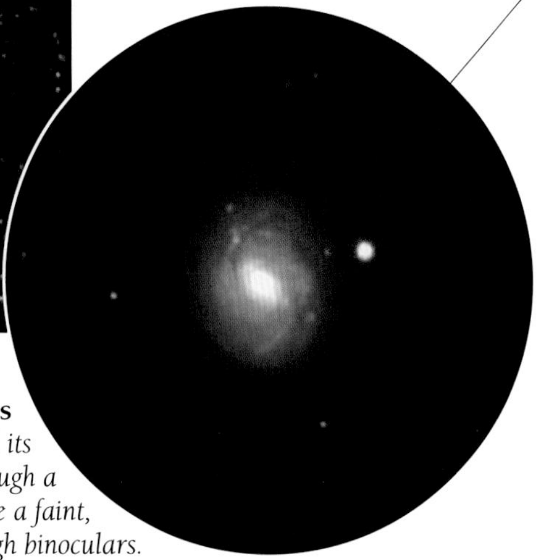

### ⓑ M77 in Cetus
*The bright center of galaxy M77 and its spiral arms can be seen clearly through a telescope. It is mag.9 and looks like a faint, circular, nebulous patch through binoculars.*

Star map labels: Perseus, Algol, California Nebula, Triangulum, Aries, Taurus, Cetus, M77, Eridanus, Fornax, Horologium

KEY TO DEEP-SKY OBJECTS

| | The Milky Way | Galaxy | Globular Cluster | Open Cluster | Diffuse Nebula | Planetary Nebula |

Andromeda

Andromeda Galaxy

Pinwheel Galaxy

Pisces

Cetus

253

288

Sculptor

Phoenix

### ✦ The Andromeda Galaxy

*The beautiful Andromeda Galaxy, M31, is a giant spiral galaxy. It can be seen with the naked eye, although its stars are not bright. However, when viewed through a telescope or in a long-exposure photograph, its spiral shape becomes clear. Andromeda has two small companions, elliptical galaxy M110 on the right of this photograph, and elliptical galaxy M32 on the left below.*

### ✦ The Pinwheel Galaxy in Triangulum

*This long-exposure photograph shows the detail and extent of the Pinwheel Galaxy, M33. The bright nucleus is formed from a dense concentration of stars. From this nucleus, arms of stars and gas spiral out. The bright dots are tight knots of gas and stars in the arms.*

### ◉ NGC 253 and NGC 288 in Sculptor

*Observers south of 50°N can see the constellation Sculptor, but its best sights, spiral galaxy NGC 253 (right) and globular cluster NGC 288 (left) can be seen at best advantage from the southern hemisphere.*

## TIPS FOR OBSERVERS

**The Pinwheel Galaxy, M33** is said to be mag.6.5. This magnitude figure represents the galaxy's total light shining as a single star. Yet its light is spread over 1° of sky, so wide-field binoculars or a telescope with an eye-piece of low magnification are best for viewing it.
**Cetus** is traditionally thought of as a whale, but some observers may see it more like a reclining chair.

| KEY TO STAR MAGNITUDE | ● –1 | ● 0 | ● 1 | ● 2 | ● 3 | ● 4 | • 5 |

JANUARY FEBRUARY MARCH APRIL MAY JUNE JULY AUGUST SEPTEMBER OCTOBER **NOVEMBER DECEMBER**

# NOVEMBER AND DECEMBER: Key Constellations

Neighboring Andromeda and Triangulum are both well-known constellations. They are not known for bright stars but for a deep-sky object that lies within their borders. In Andromeda is the beautiful Andromeda Galaxy; Triangulum has the lovely Pinwheel. For most observers, the Andromeda Galaxy is the most distant object visible with the naked eye. It is a neighboring galaxy to our own galaxy, the Milky Way.

## ANDROMEDA *Andromeda*

Width

Andromeda's stars do not fall into an obvious pattern and the best way of finding the constellation is by locating the Square of Pegasus (p.128) with its four bright stars, one of which, Alpheratz, is part of Andromeda. A line can be drawn from this star to Mirach. A short way from Mirach is the Andromeda Galaxy (M31); it can be seen with the naked eye as a faint blur of light.

*Gamma (γ) is a double star.*

*Galaxies M32 and M110 lie close to the Andromeda Galaxy.*

Andromeda Galaxy
M110
M32
Almach
Mirach
Alpheratz

*This planetary nebula can be viewed only through a telescope.*

⊙ **The Andromeda Galaxy**
*Binoculars reveal the spiral shape of the Andromeda Galaxy. It has a distinctive, star-packed nucleus that shows up as a bright center within a glow of light.*

## ERIDANUS *The River Eridanus*

Width

Eridanus flows across the sky linking two bright stars: Rigel in Orion (p.112) at the top and Achernar (in Eridanus) at the bottom. Achernar is the 9th brightest star in the sky and its name is Arabic for "end of the river." Between these two stars is a river of fainter stars. As the river flows south, starting very near to the celestial equator, to almost 60°S, many observers will only see part of its path.

*This planetary nebula looks like a pale, blue star when seen through binoculars.*

*This is a barred spiral galaxy.*

Achernar

⊙ **Stars in Eridanus**
*The path of Eridanus can be seen with the naked eye. This is the southern tip with Achernar just above the horizon; Chi (χ) and Phi (φ) are below, and just to the left of center.*

*A blue-white star of mag.0.5, Achernar marks the constellation's southernmost tip.*

| KEY TO DEEP-SKY OBJECTS |  The Milky Way |  Galaxy |  Globular Cluster | Open Cluster |  Diffuse Nebula |  Planetary Nebula |

## TRIANGULUM *The Triangle*

Width 🖐

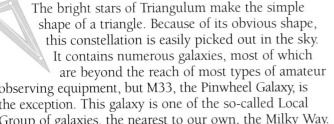

The bright stars of Triangulum make the simple shape of a triangle. Because of its obvious shape, this constellation is easily picked out in the sky. It contains numerous galaxies, most of which are beyond the reach of most types of amateur observing equipment, but M33, the Pinwheel Galaxy, is the exception. This galaxy is one of the so-called Local Group of galaxies, the nearest to our own, the Milky Way.

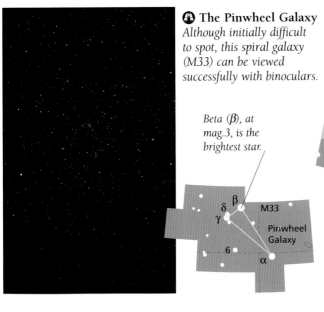

🔭 **The Pinwheel Galaxy**
*Although initially difficult to spot, this spiral galaxy (M33) can be viewed successfully with binoculars.*

*Beta (β), at mag.3, is the brightest star.*

## PERSEUS *Perseus*

Width 🖐

Perseus was the hero of Greek mythology who decapitated the gorgon, Medusa. The bright star Algol, representing Medusa's evil eye, is an eclipsing binary (p.87), that varies in brightness from mag.2.1 to mag.3.5 in just under three days. Perseus is also the constellation from which the Perseid meteor shower (pp.74–5) radiates in August.

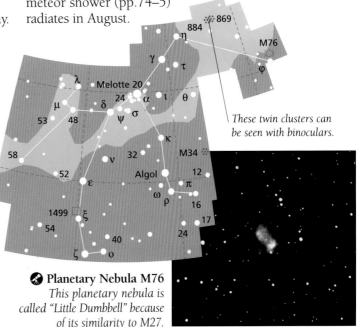

*These twin clusters can be seen with binoculars.*

💫 **Planetary Nebula M76**
*This planetary nebula is called "Little Dumbbell" because of its similarity to M27.*

## CETUS *The Whale*

Width 🖐🖐

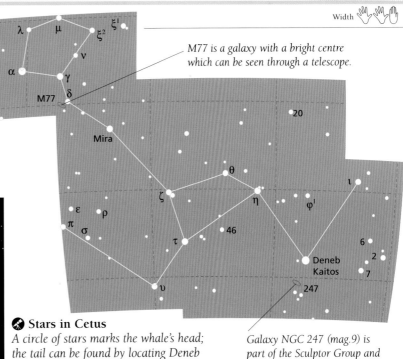

Cetus is the fourth-largest constellation in the sky. Its head is in the north celestial sky and its body in the south. It has no bright stars and the only star of any note is Mira, a pulsating red giant. Most of the time this variable star (p.87) cannot be seen with the naked eye, but at its brightest it reaches mag.3.

*M77 is a galaxy with a bright centre which can be seen through a telescope.*

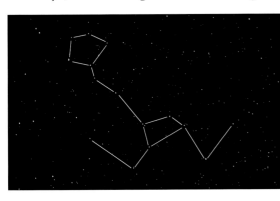

🔭 **Stars in Cetus**
*A circle of stars marks the whale's head; the tail can be found by locating Deneb Kaitos, the bright star at bottom right.*

*Galaxy NGC 247 (mag.9) is part of the Sculptor Group and can be seen with a telescope.*

| KEY TO STAR MAGNITUDE | ⬤ –1 | ⬤ 0 | ⬤ 1 | ⬤ 2 | ⚬ 3 | • 4 | · 5 | · 6 |

JANUARY FEBRUARY MARCH APRIL MAY JUNE JULY AUGUST SEPTEMBER OCTOBER NOVEMBER DECEMBER

133

# DIARY OF EVENTS

This month-by-month guide highlights objects and events seen only, or at their best, at certain times of the year. The guide can be used for any year, and in northern and southern latitudes. Set the planisphere to see if, and at what time, objects are in your sky, and look at the sunrise and sunset tables to see when your sky is dark.

**Preparing a telescope**

## JANUARY

### MAIN FEATURES

**M42**, the Orion Nebula in the sword of Orion the Hunter (p.112), is a prominent nebula best seen with the naked eye when it is high above the horizon. **The Hyades and Pleiades** (Taurus) can be seen with the naked eye.

### ALSO VISIBLE

Aldebaran (Taurus); Castor and Pollux (Gemini); Sirius (Canis Major); Procyon (Canis Minor); M36, M37, and M38 (Auriga); M41 (Canis Major); M44 (Cancer)

### METEOR SHOWER

Northern hemisphere observers can see the Quadrantids around January 3 and 4. They radiate from the northern part of the constellation, Boötes (p.120). The shower is best seen after midnight.

## FEBRUARY

### MAIN FEATURES

**M41** (Canis Major) can be seen with the naked eye in good observing conditions. **NGC 2244** is a star cluster seen at the center of the Rosette Nebula (Monoceros). **Betelgeuse** (Orion), **Procyon** (Canis Minor), and **Sirius** (Canis Major) are three brilliant stars that form the so-called Winter Triangle for northern hemisphere observers. The same star triangle can also be seen from the southern hemisphere where it is summer.

### ALSO VISIBLE

Aldebaran (Taurus); Castor and Pollux (Gemini); Rigel (Orion); M36, M37, and M38 (Auriga); M44 – the Beehive Cluster (Cancer); the Hyades and Pleiades star clusters (Taurus); M42, the Orion Nebula (Orion); NGC 3372, the Eta Carinae Nebula (Carina)

## MARCH

### MAIN FEATURES

**M44**, the Beehive Cluster (Cancer), can be seen with the naked eye in good observing conditions. Through binoculars a swarm of faint stars becomes visible. **NGC 3372**, the Eta Carinae Nebula (Carina), lies along the path of the Milky Way and looks like a bright patch to the naked eye. **The Coalsack** (Crux) is a dark nebula in the path of the Milk Way. It looks like a hole in the sky.

### ALSO VISIBLE

Castor and Pollux (Gemini); Rigel and Betelgeuse (Orion); Sirius (Canis Major); Procyon (Canis Minor); Regulus (Leo); Spica (Virgo); M36, M37, and M38 (Auriga); M41 (Canis Major); NGC 4755 (Crux); M42, Orion Nebula (Orion); NGC 2244 (Monoceros)

## JULY

### MAIN FEATURES

**M22** (Sagittarius) is one of the best globular clusters to be seen in the sky. It looks very much like a fuzzy star to the naked eye, but the view is much improved through binoculars. A triangle of stars, commonly called the **Summer Triangle** by northern hemisphere observers, consists of three bright stars positioned in different constellations—**Vega** (Lyra), **Deneb** (Cygnus), and **Altair** (Aquila). **The Cygnus Rift** (Cygnus), sometimes known as the Northern Coalsack, is a dark nebula (pp.88–9) positioned across the path of the Milky Way.

### ALSO VISIBLE

Arcturus in Boötes; Spica (Virgo); Antares (Scorpius); M13 (Hercules); M6 and M7 (Scorpius); M27, planetary nebula (Vulpecula)

## AUGUST

### MAIN FEATURES

**M8**, the Lagoon Nebula (Sagittarius), can be spotted with the naked eye in a dark sky. The nebula's name comes from the dark dust lane over its center. Using binoculars, a swarm of faint stars is seen. **Albireo** (Cygnus) is a double star that is easily resolved with a small telescope.

### ALSO VISIBLE

Vega (Lyra); Deneb (Cygnus); Altair (Aquila); Antares (Scorpius); Fomalhaut (Piscis Austrinus); M13 (Hercules); M22 (Sagittarius); NGC 104 (Tucana); M27, planetary nebula (Vulpecula)

### METEOR SHOWER

The Perseids is the richest meteor shower. It peaks in the middle of August but can be seen for much of the month.

## SEPTEMBER

### MAIN FEATURES

**M27**, a planetary nebula called the Dumbbell Nebula because of its shape, can be spotted through binoculars and its shape seen clearly in a telescope. It makes a good subject for CCD imaging. **Delta Cephei** is a variable star that changes regularly in brightness. A small telescope shows that it also has a fainter companion star. **NGC 104** (Tucana) is the second-best globular cluster in the sky. It is also known as 47 Tucanae, and looks very much like a fuzzy star to the naked eye.

### ALSO VISIBLE

Vega (Lyra); Deneb (Cygnus); Altair (Aquila); Fomalhaut (Piscis Austrinus); NGC 869 and NGC 884, the Double Cluster (Perseus); M31, the Andromeda Galaxy (Andromeda); the Cygnus Rift (Cygnus)

Observing the night sky with binoculars

## APRIL

### MAIN FEATURES

**The Jewel Box Cluster**, NGC 4755 (Crux), is high in the late-evening sky for southern observers. It is visible to the naked eye as a fuzzy star, while through binoculars or a telescope its different-colored stars shine like jewels.

### ALSO VISIBLE

Arcturus (Boötes); Spica (Virgo); Regulus (Leo); Procyon (Canis Minor); M44 (Cancer); NGC 2244 (Monoceros); NGC 5139 Omega Centauri (Centaurus); the Coalsack (Crux); NGC 3372, the Eta Carinae Nebula (Carina)

### METEOR SHOWER

Northern observers can see the Lyrids around April 21. They radiate from a point near the bright star, Vega, in Lyra.

## MAY

### MAIN FEATURES

**Omega Centauri**, NGC 5139, is the best globular cluster in the night sky. It is well placed in the late evening for southern observers, and is clearly seen as a fuzzy star with the naked eye.

### ALSO VISIBLE

Arcturus (Boötes); Spica (Virgo); Regulus (Leo); Antares (Scorpius); Vega (Lyra); M6 and M7 (Scorpius); M44 (Cancer); NGC 4755 (Crux); M13 (Hercules); NGC 3372 the Eta Carinae Nebula (Carina)

### METEOR SHOWER

Observers in the southern hemisphere can see the Eta Aquarids. They radiate from the constellation Aquarius in the last few days of April through to late May, peaking at around May 5.

## JUNE

### MAIN FEATURES

**M13** in Hercules (p.124), the most spectacular globular cluster in the northern sky, is positioned high above the horizon in the late evening. It can be seen clearly with the naked eye against a dark background.
**M6 and M7** are two impressive star clusters in Scorpius (p.109). They can both be seen with the naked eye against the backdrop of the Milky Way, but are best viewed through binoculars with their wider field of view when many of the stars in the clusters can be seen.

### ALSO VISIBLE

Arcturus (Boötes); Spica (Virgo); Antares (Scorpius); Vega (Lyra); (Sagittarius); NGC 5139 Omega Centauri (Centaurus); M8, the Lagoon Nebula (Sagittarius); M27, planetary nebula (Vulpecula)

## OCTOBER

### MAIN FEATURES

**M31**, the distant Andromeda Galaxy (Andromeda) can be seen by naked eye. The Great Square of Pegasus is formed by three stars in Pegasus, and one in Andromeda, well placed in the sky for northern and southern observers.

### ALSO VISIBLE

Fomalhaut (Piscis Austrinius); Achernar (Eridanus); NGC 104 (Tucana); the Hyades and Pleiades (Taurus); NGC 869 and NGC 884 (Perseus); the Cygnus Rift (Cygnus)

### METEOR SHOWER

The Orionids are seen in the second half of the month. They radiate from a point in Orion (p.112), and are best seen after midnight on or about October 21.

## NOVEMBER

### MAIN FEATURES

**NGC 869** and **NGC 884** (Perseus), the Double Cluster, lies in a spiral arm of the Milky Way. They can be seen by naked eye but binoculars give a clearer view.

### ALSO VISIBLE

Betelgeuse and Rigel (Orion); Aldebaran (Taurus); Capella (Auriga); Achenar (Eridanus); the Hyades and Pleaides (Taurus); M36, M37, and M38 (Auriga); M41 (Canis Major); NGC 2070, the Tarantula Nebula (Dorado); M42, the Orion Nebula (Orion); M31, the Andromeda Galaxy

### METEOR SHOWERS

The Taurids peak in early November but can be seen from late October through November. The Leonids radiating from Leo peak around November 17.

## DECEMBER

### MAIN FEATURES

**M36, M37**, and **M38** (Auriga) can be seen through binoculars. A telescope shows the stars in the clusters. The Large Magellanic Cloud (Dorado) is a companion galaxy to the Milky Way. It is a fuzzy patch to the naked eye but can be seen clearly through binoculars.

### ALSO VISIBLE

Betelgeuse and Rigel (Orion); Aldebaran (Taurus); Capella (Auriga); Achernar (Eridanus); Procyon (Canis Minor); Canopus (Carina); M41 (Canis Major); M44, the Beehive Cluster (Cancer); the Orion Nebula

### METEOR SHOWER

The Germinids radiate from near the bright star, Castor (Gemini). This rich shower is at its peak around December 13.

# ASTRONOMICAL DATA

## THE CONSTELLATIONS

*The celestial sphere is divided into 88 constellations. These have been agreed by the International Astronomical Union. This table lists all 88, with Latin, common names, and genitive, denoting "belonging to."*

| Name | Genitive | Common name |
|---|---|---|
| Andromeda | Andromedae | Andromeda |
| Antlia | Antliae | The Air Pump |
| Apus | Apodis | The Bird of Paradise |
| Aquarius | Aquarii | The Water Carrier |
| Aquila | Aquilae | The Eagle |
| Ara | Arae | The Altar |
| Aries | Arietis | The Ram |
| Auriga | Aurigae | The Charioteer |
| Boötes | Boötis | The Herdsman |
| Caelum | Caeli | The Chisel |
| Camelopardalis | Camelopardalis | The Giraffe |
| Cancer | Cancri | The Crab |
| Canes Venatici | Canum Venaticorum | The Hunting Dogs |
| Canis Major | Canis Majoris | The Great Dog |
| Canis Minor | Canis Minoris | The Little Dog |
| Capricornus | Capricorni | The Sea Goat |
| Carina | Carinae | The Keel |
| Cassiopeia | Cassiopeiae | Cassiopeia |
| Centaurus | Centauri | The Centaur |
| Cepheus | Cephei | Cepheus |
| Cetus | Ceti | The Whale |
| Chemaeleon | Chamaeleonis | The Chameleon |
| Circinus | Circini | The Compass |
| Columba | Columbae | The Dove |
| Coma Berenices | Coma Berenicis | Berenice's Hair |
| Corona Australis | Coronae Australis | The Southern Crown |
| Corona Borealis | Coronae Borealis | The Northern Crown |
| Corvus | Corvi | The Crow |
| Crater | Crateris | The Cup |
| Crux | Crucis | The Southern Cross |
| Cygnus | Cygni | The Swan |
| Delphinus | Delphini | The Dolphin |
| Dorado | Doradus | The Goldfish |
| Draco | Draconis | The Dragon |
| Equuleus | Equulei | The Little Horse |
| Eridanus | Eridani | The River Eridanus |
| Fornax | Fornacis | The Furnace |
| Gemini | Geminorum | The Twins |
| Grus | Gruis | The Crane |
| Hercules | Herculis | Hercules |
| Horologium | Horologii | The Pendulum Clock |

| Name | Genitive | Common name |
|---|---|---|
| Hydra | Hydrae | The Water Snake |
| Hydrus | Hydri | The Little Water Snake |
| Indus | Indi | The Indian |
| Lacerta | Lacertae | The Lizard |
| Leo | Leonis | The Lion |
| Leo Minor | Leonis Minoris | The Little Lion |
| Lepus | Leporis | The Hare |
| Libra | Librae | The Scales |
| Lupus | Lupi | The Wolf |
| Lynx | Lyncis | The Lynx |
| Lyra | Lyrae | The Lyre |
| Mensa | Mensae | Table Mountain |
| Microscopium | Microscopii | The Microscope |
| Monoceros | Monocerotis | The Unicorn |
| Musca | Muscae | The Fly |
| Norma | Normae | The Level |
| Octans | Octantis | The Octant |
| Ophiuchus | Ophiuchi | The Serpent Bearer |
| Orion | Orionis | The Hunter |
| Pavo | Pavonis | The Peacock |
| Pegasus | Pegasi | Pegasus |
| Perseus | Persei | Perseus |
| Phoenix | Phoenicis | The Phoenix |
| Pictor | Pictoris | The Painter's Easel |
| Pisces | Piscium | The Fishes |
| Piscis Austrinus | Piscis Austrini | The Southern Fish |
| Puppis | Puppis | The Stern |
| Pyxis | Pyxidis | The Mariner's Compass |
| Reticulum | Reticuli | The Net |
| Sagitta | Sagittae | The Arrow |
| Sagittarius | Sagittarii | The Archer |
| Scorpius | Scorpii | The Scorpion |
| Sculptor | Sculptoris | The Sculptor |
| Scutum | Scuti | The Shield |
| Serpens | Serpentis | The Serpent |
| Sextans | Sextantis | The Sextant |
| Taurus | Tauri | The Bull |
| Telescopium | Telescopii | The Telescope |
| Triangulum | Trianguli | The Triangle |
| Triangulum Australe | Trianguli Australis | The Southern Triangle |
| Tucana | Tucanae | The Toucan |
| Ursa Major | Ursae Majoris | The Great Bear |
| Ursa Minor | Ursae Minoris | The Little Bear |
| Vela | Velorum | The Sail |
| Virgo | Virginis | The Virgin |
| Volans | Volantis | The Flying Fish |
| Vulpecula | Vulpeculae | The Fox |

## SUNRISE AND SUNSET TIMES

| Latitude | JANUARY 15 Rise | Set | FEBRUARY 15 Rise | Set | MARCH 15 Rise | Set | APRIL 15 Rise | Set | MAY 15 Rise | Set | JUNE 15 Rise | Set |
|---|---|---|---|---|---|---|---|---|---|---|---|---|
| 60°N | 08.50 | 15.30 | 07.40 | 16.50 | 06.20 | 18.00 | 04.40 | 19.20 | 03.20 | 20.30 | 02.40 | 21.20 |
| 40°N | 07.20 | 17.00 | 06.50 | 17.40 | 06.10 | 18.10 | 05.20 | 18.40 | 04.40 | 19.10 | 04.30 | 19.30 |
| 20°N | 06.40 | 17.40 | 06.30 | 18.00 | 06.10 | 18.10 | 05.40 | 18.20 | 05.20 | 18.30 | 05.20 | 18.40 |
| 0° | 06.10 | 18.10 | 06.10 | 18.20 | 06.10 | 18.10 | 06.00 | 18.00 | 05.50 | 18.00 | 06.00 | 18.00 |
| 20°S | 05.30 | 18.50 | 05.50 | 18.40 | 06.00 | 18.20 | 06.10 | 17.50 | 06.20 | 17.30 | 06.30 | 17.30 |
| 40°S | 04.50 | 19.30 | 05.30 | 19.00 | 06.00 | 18.20 | 06.30 | 17.30 | 07.00 | 17.00 | 07.20 | 16.40 |

| Latitude | JULY 15 Rise | Set | AUGUST 15 Rise | Set | SEPTEMBER 15 Rise | Set | OCTOBER 15 Rise | Set | NOVEMBER 15 Rise | Set | DECEMBER 15 Rise | Set |
|---|---|---|---|---|---|---|---|---|---|---|---|---|
| 60°N | 03.00 | 21.10 | 04.10 | 19.50 | 05.30 | 18.20 | 06.40 | 16.50 | 08.00 | 15.30 | 09.00 | 14.50 |
| 40°N | 04.40 | 19.30 | 05.10 | 17.00 | 05.40 | 18.10 | 06.10 | 17.20 | 06.50 | 16.40 | 07.10 | 16.40 |
| 20°N | 05.30 | 18.40 | 05.40 | 18.30 | 05.50 | 18.00 | 05.50 | 17.40 | 06.10 | 17.20 | 06.30 | 17.20 |
| 0° | 06.00 | 18.10 | 06.00 | 18.10 | 05.50 | 18.00 | 05.40 | 17.50 | 05.40 | 17.50 | 05.50 | 18.00 |
| 20°S | 06.40 | 17.40 | 06.20 | 17.50 | 06.00 | 18.00 | 05.30 | 18.00 | 05.10 | 18.20 | 05.20 | 18.40 |
| 40°S | 07.20 | 16.50 | 06.50 | 17.20 | 06.00 | 17.50 | 05.10 | 18.20 | 04.40 | 18.50 | 04.30 | 19.20 |

## THE MESSIER OBJECTS

*Many of the stellar objects mentioned in this book are Messier objects. They are identified by the letter "M," followed by a number. These numbers were given to just over 100 objects by the French astronomer, Charles Messier, in the late 18th century. He listed the objects to avoid confusion with comets. Today, Messier's work is used by astronomers as a catalog of the best galaxies and star clusters in the night sky.*

| No | Constellation | Description |
|----|---------------|-------------|
| 1 | Taurus | Supernova remnant |
| 2 | Aquarius | Globular cluster |
| 3 | Canes Vanetici | Globular cluster |
| 4 | Scorpius | Globular cluster |
| 5 | Serpens | Globular cluster |
| 6 | Scorpius | Open cluster |
| 7 | Scorpius | Open cluster |
| 8 | Sagittarius | Diffuse nebula |
| 9 | Ophiuchus | Globular cluster |
| 10 | Ophiuchus | Globular cluster |
| 11 | Scutum | Open cluster |
| 12 | Ophiuchus | Globular cluster |
| 13 | Hercules | Globular cluster |
| 14 | Ophiuchus | Globular cluster |
| 15 | Pegasus | Globular cluster |
| 16 | Serpens | Open cluster |
| 17 | Sagittarius | Diffuse nebula |
| 18 | Sagittarius | Open cluster |
| 19 | Ophiuchus | Globular cluster |
| 20 | Sagittarius | Diffuse nebula |
| 21 | Sagittarius | Open cluster |
| 22 | Sagittarius | Globular cluster |
| 23 | Sagittarius | Open cluster |
| 24 | Sagittarius | Star field |
| 25 | Sagittarius | Open cluster |
| 26 | Scutum | Open cluster |

| No | Constellation | Description |
|----|---------------|-------------|
| 27 | Vulpecula | Planetary nebula |
| 28 | Sagittarius | Globular cluster |
| 29 | Cygnus | Open cluster |
| 30 | Capricornus | Globular cluster |
| 31 | Andromeda | Spiral galaxy |
| 32 | Andromeda | Elliptical galaxy |
| 33 | Triangulum | Spiral galaxy |
| 34 | Perseus | Open cluster |
| 35 | Gemini | Open cluster |
| 36 | Auriga | Open cluster |
| 37 | Auriga | Open cluster |
| 38 | Auriga | Open cluster |
| 39 | Cygnus | Open cluster |
| 40 | Ursa Major | Faint double star |
| 41 | Canis Major | Open cluster |
| 42 | Orion | Diffuse nebula |
| 43 | Orion | Diffuse nebula |
| 44 | Cancer | Open cluster |
| 45 | Taurus | Open cluster |
| 46 | Puppis | Open cluster |
| 47 | Puppis | Open cluster |
| 48 | Hyades | Open cluster |
| 49 | Virgo | Elliptical galaxy |
| 50 | Monoceros | Open cluster |
| 51 | Canes Venatici | Spiral galaxy |
| 52 | Cassiopeia | Open cluster |
| 53 | Coma Berenices | Globular cluster |
| 54 | Sagittarius | Globular cluster |
| 55 | Sagittarius | Globular cluster |
| 56 | Lyra | Globular cluster |
| 57 | Lyra | Planetary nebula |
| 58 | Virgo | Spiral galaxy |
| 59 | Virgo | Elliptical galaxy |
| 60 | Virgo | Elliptical galaxy |
| 61 | Virgo | Spiral galaxy |
| 62 | Ophiuchus | Globular cluster |
| 63 | Canes Venatici | Spiral galaxy |
| 64 | Coma Berenices | Spiral galaxy |
| 65 | Leo | Spiral galaxy |
| 66 | Leo | Spiral galaxy |
| 67 | Cancer | Open cluster |
| 68 | Hydra | Globular cluster |

| No | Constellation | Description |
|----|---------------|-------------|
| 69 | Sagittarius | Globular cluster |
| 70 | Sagittarius | Globular cluster |
| 71 | Sagitta | Globular cluster |
| 72 | Aquarius | Globular cluster |
| 73 | Aquarius | Small star goup |
| 74 | Pisces | Spiral galaxy |
| 75 | Sagittarius | Globular cluster |
| 76 | Perseus | Planetary nebula |
| 77 | Cetus | Spiral galaxy |
| 78 | Orion | Diffuse nebula |
| 79 | Lepus | Globular cluster |
| 80 | Scorpius | Globular cluster |
| 81 | Ursa Major | Spiral galaxy |
| 82 | Ursa Major | Irregular galaxy |
| 83 | Hyades | Spiral galaxy |
| 84 | Virgo | Elliptical galaxy |
| 85 | Coma Berenices | Elliptical galaxy |
| 86 | Virgo | Elliptical galaxy |
| 87 | Virgo | Elliptical galaxy |
| 88 | Coma Berenices | Spiral galaxy |
| 89 | Virgo | Elliptical galaxy |
| 90 | Virgo | Spiral galaxy |
| 91 | Coma Berenices | Spiral galaxy |
| 92 | Hercules | Globular cluster |
| 93 | Puppis | Open cluster |
| 94 | Canes Venatici | Spiral galaxy |
| 95 | Leo | Spiral galaxy |
| 96 | Leo | Spiral galaxy |
| 97 | Ursa Major | Planetary nebula |
| 98 | Coma Berenices | Spiral galaxy |
| 99 | Coma Berenices | Spiral galaxy |
| 100 | Coma Berenices | Spiral galaxy |
| 101 | Ursa Major | Spiral galaxy |
| 102 | Duplicate of M101, above | |
| 103 | Cassiopeia | Open cluster |
| 104 | Virgo | Spiral galaxy |
| 105 | Leo | Elliptical galaxy |
| 106 | Canes Venatici | Spiral galaxy |
| 107 | Ophiuchus | Globular cluster |
| 108 | Ursa Major | Spiral galaxy |
| 109 | Ursa Major | Spiral galaxy |
| 110 | Andomeda | Elliptical galaxy |

## COMETS 1999 TO 2010

| Name of periodic comet | Observing date | Returns (years) | Constellation where best | Magnitude |
|------------------------|----------------|-----------------|--------------------------|-----------|
| Tempel 2 | 09.1999 | 5.47 | Scorpius | 8.5 |
| Schwassmann-Wachmann 3 | 01.2001 | 5.36 | Ophiuchus | 7.0 |
| Honda-Mrkos-Pajdusakova | 03.2001 | 5.25 | Aries | 9.0 |
| Borelly | 09.2001 | 6.86 | Cancer | 9.5 |
| Encke | 12.2003 | 3.30 | Aquila | 7.0 |
| Tempel 1 | 06.2005 | 5.51 | Virgo | 9.5 |
| Chernykh | 10.2005 | 13.9 | Cetus | 10 |
| Schwassmann-Wachmann 3 | 05.2006 | 5.36 | Hercules | 1.5 |
| Honda-Mrkos-Pajdusakova | 06.2006 | 5.25 | Aries | 10 |
| Faye | 11.2006 | 7.55 | Pisces | 10 |
| Tuttle | 01.2008 | 13.6 | Pisces | 5.0 |
| Kopff | 07.2009 | 6.44 | Aquarius | 9.0 |
| Wild 2 | 03.2010 | 6.42 | Virgo | 8.5 |
| Tempel 2 | 07.2010 | 5.37 | Cetus | 8.0 |
| Hartley | 10.2010 | 6.47 | Gemini | 5.0 |

## PLANETARY MOONS OVER 1,900 MILES IN DIAMETER

| Moons | Planets | Average Distance from planet (miles) | Orbital Period days | Diameter miles | Brightest magnitude |
|-------|---------|--------------------------------------|---------------------|----------------|---------------------|
| Ganymede | Jupiter | 664,866 | 7.2 | 3,270 | 4.6 |
| Titan | Saturn | 759,190 | 15.9 | 3,200 | 8.4 |
| Callisto | Jupiter | 1,170,040 | 16.7 | 2,982 | 5.6 |
| Io | Jupiter | 262,218 | 1.8 | 2,255 | 5.0 |
| Moon | Earth | 238,917 | 27.3 | 2,160 | −12.5 |
| Europa | Jupiter | 416,877 | 3.5 | 1,950 | 5.3 |

## CONVERTING UNIVERSAL TIME TO LOCAL TIME

The time of an event is given in Universal Time (U.T.). This is the system used for astronomical events. It is based on the time at 0 longitude, Greenwich, England. To convert UT to your local time, you must first know which time zone you live in. A time zone is usually 15° of longitude wide and equal to one hour difference in time. Add or subtract your time zone difference from U.T. to get the local time of an event. (Add if east of Greenwich or subtract if west of Greenwich. Remember to adjust your figure by one hour if your location is using daylight saving time on the date of observation.

The planisphere (pp.22–3) is based on local time, and the sunrise and sunset figures (opposite) are also in local time. The only adjustment needed is to add one hour for daylight saving if this applies to you.

# GLOSSARY

**ABSOLUTE MAGNITUDE**
A measure of the luminosity of an object that exists in space, defined as how bright the object would seem if it were at a standard distance from Earth.

**APERTURE**
The opening through which light enters an astronomical instrument and the diameter of the main lens or mirror in a telescope.

**APPARENT MAGNITUDE**
A measure of the brightness of an object in space as it is perceived by an observer on Earth. *See also* Magnitude.

**ASTERISM**
A prominent pattern created by stars in one or more constellations—for example, the Big Dipper, that is part of the constellation Ursa Major and the Great Square of Pegasus, formed by the stars of Pegasus and Andromeda.

**ASTEROID**
A small rocky object orbiting the Sun.

**AXIS**
An imaginary line through a celestial object around which it rotates.

**BINARY STAR**
Two stars that are linked by gravity, orbiting around a common center of mass. *See also* Eclipsing binary.

**CELESTIAL EQUATOR**
An imaginary circle that divides the celestial sphere into the northern and southern celestial hemispheres. It is the celestial extension of Earth's equator.

**CELESTIAL SPHERE**
An imaginary sphere of immense size surrounding Earth on which celestial objects appear to be positioned.

**CEPHEID VARIABLE**
A type of variable star that pulsates regularly in size and brightness, every few days or weeks with a period linked to its average luminosity.

**CHARGE-COUPLED DEVICE (CCD)**
A light-sensitive electronic detector used in place of photographic film to record images of celestial objects. It is extremely responsive to light, and can therefore detect even objects that are only faintly illuminated as well as bright objects.

**CIRCUMPOLAR STAR**
A star that remains above the horizon without rising or setting, as observed from a particular location on Earth. The star can be seen to circle the north or south celestial poles.

**COMET**
A mountain-sized body of snow and dust orbiting the Sun. A comet produces a head and tails as it travels near the Sun.

**CONJUNCTION**
An occasion when two bodies in the Solar System (such as the Sun and a planet) have the same longitude as seen from Earth, and appear to be close in the sky.

**CONSTELLATION**
An area of sky, originally a star pattern, but now defined as the area within boundaries set by the International Astronomical Union. There are 88 constellations within the celestial sphere.

**DARK NEBULA**
A cloud of interstellar gas and dust that obscures light from stars or other objects positioned behind it. The Horsehead Nebula in Orion is an example.

**DECLINATION**
A coordinate on the celestial sphere. The equivalent of latitude on Earth, it is the angle between an object and the celestial equator. Declination is measured north or south of the celestial equator.

**DEEP-SKY OBJECT**
Objects beyond the Solar System, such as star clusters, nebulae, and galaxies.

**DIRECT MOTION**
The usual motion of objects in the Solar System that occurs in one of three ways:

1 The apparent west to east motion of an object as it is observed from Earth against the background of stars.

2 The counterclockwise orbital motion (observed from above an object's north pole) of a body traveling around the Sun, or a moon around a planet.

3 The counterclockwise spin of a body (seen from above its north pole).

**DOUBLE STAR**
Two stars that appear to be close together as observed from Earth. *See also* Binary star, Optical double Star.

**ECLIPSE**
An occasion when one celestial object is hidden from view by another celestial object, as seen from Earth.

**ECLIPSING BINARY**
A pair of stars that are linked by gravity. They are known as an eclipsing binary because one star periodically passes in front of the other as seen from Earth, temporarily cutting off its light.

**ECLIPTIC**
The plane of Earth's orbit of the Sun, projected onto the celestial sphere. The Sun appears to move along the ecliptic across Earth's sky each year. The planets appear close to the ecliptic.

**ELONGATION**
The angle between a planet and the Sun, or between a moon and its accompanying planet, as viewed from Earth.

**EXTRAGALACTIC**
A term for other galaxies beyond the reaches of our own galaxy, the Milky Way.

**GALAXY**
A mass of stars, numbering many millions all held together by gravity.

**GIANT STAR**
A larger, luminous star in the late stages of its development. Giant stars are at least 10 times the diameter of the Sun and a thousand times as luminous.

**GLOBULAR CLUSTER**
A ball-shaped, densely packed group of stars, containing tens to hundreds of thousands of members. Globular clusters contain some of the oldest stars known.

**LIGHT YEAR**
A unit of distance, not time. It is the distance a beam of light covers in a year, or 5,878,600,000,000 miles (9,460,700,000,000 km).

**LUMINOSITY**
The intrinsic or absolute amount of energy radiated per second by a celestial object. The unit of measurement is termed absolute magnitude.

**MAGNITUDE**
The brightness of a celestial object, measured on a numerical scale with brighter objects given small or negative values and faint objects given high values. *See also* Absolute magnitude, Apparent magnitude.

**METEOR**
A streak of light in the sky, caused by a small particle of dust or a piece of

rock from space entering Earth's upper atmosphere and burning up.

**METEORITE**
A fragment of an asteroid, planet, or moon that has fallen through space to land on the surface of a planet or satellite.

**MILKY WAY**
The faint, hazy band of light that can be observed with the naked eye crossing the sky on dark nights, that is composed of countless distant stars in our own Galaxy. It is also a popular name for our Galaxy as a whole.

**MIRA VARIABLE**
A red giant star that varies in size and brightness over a period of months or years, varying in brightness by several magnitudes

**NEBULA**
A cloud of gas and dust, usually found in the spiral arms of a galaxy. Some nebulae are bright, being lit up by stars within them, while others are dark, and can be seen only if silhouetted against a brighter background. See also Planetary Nebula.

**NOVAE**
A star that erupts in brightness by about 10 magnitudes for a period of time, before declining again. Novae occur in close binaries, one member of which is a white dwarf. Gas flows from the companion star onto the white dwarf, causing an explosion that leads to the brightening.

**OPEN CLUSTER**
An irregularly-shaped, loose grouping of dozens or hundreds of relatively young stars, usually found in the spiral arms of a galaxy. The stars are spaced much further apart than those in a globular cluster.

**OPPOSITION**
The occasion when a body in the Solar System appears opposite the Sun in the sky as seen from Earth and is therefore visible all night.

**OPTICAL DOUBLE STAR**
Two stars that appear to be positioned close to one another in the sky but in fact lie at different distances from Earth.

**PARALLAX**
The change in the apparent position of an object when it is seen from two different locations. How much the object's position appears to change depends on its distance from Earth as well as the separation of the observing locations. Nearby stars show a slight parallax shift as Earth orbits the Sun; this shift is used to calculate their distances.

**PHASE**
A fraction of the illuminated disk of a planet or moon, as seen from Earth.

**PLANET**
A body orbiting the Sun or another star. A planet shines by reflected light.

**PLANETARY NEBULA**
A shell of gas thrown off by a star late in its development. Through a small telescope, the shell can resemble the disk of a planet, which is how the name arose. In fact, it takes various forms such as rings, circles, and dumbbell-like or irregular shapes.

**RADIANT**
The point in the sky from which a meteor shower appears to originate.

**REFLECTING TELESCOPE** (Reflector)
A type of telescopic instrument in which light is collected and focused by a mirror.

**REFRACTING TELESCOPE** (Refractor)
A type of telescopic instrument in which light is collected and focused by a lens.

**RESOLVE**
To distinguish astronomical detail, for example to see individual stars in a cluster, or craters on the Moon.

**RETROGRADE MOTION**
A motion that is opposite to the normal, direct motion in the Solar System.

1 The apparent movement from east to west of an object seen against the starry background from Earth.

2 The clockwise (as seen from the object's north pole) orbit of an object around the Sun, or a moon around a planet.

3 The clockwise spin of a body (as observed from its north pole).

**RIGHT ASCENSION**
A coordinate on the celestial sphere, the equivalent of longitude on Earth. It is measured eastward along the celestial equator from the point where the Sun moves north across the celestial equator (known as the vernal equinox).

**SCHMIDT-CASSEGRAIN TELESCOPE**
A type of reflecting telescope that uses a correcting lens to eliminate aberration (divergence of the light rays forming the image). This type of telescope is quite compact and therefore easily portable.

**SEEING**
The quality of observing conditions when making telescopic observations. The steadiness of the image is affected by random motion in the atmosphere.

**SOLAR SYSTEM**
The Sun and the various bodies in orbit around it, including the nine planets and their moons, as well as asteroids, comets, and pieces of debris.

**STAR**
A luminous sphere of gas that produces energy by nuclear reaction in its core. Most stars are made largely of hydrogen and helium.

**SUPERGIANT STAR**
The largest and most luminous type of star. It is the late stage in the life of a star which is at least 10 times as massive as the Sun.

**SUPERNOVA**
A massive star that explodes violently at the end of its life. The star is virtually destroyed but the explosion makes it shine brightly for a few weeks or months.

**SUPERNOVA REMNANT**
The expanding shell of gas that is blown out by the explosion of a supernova.

**UNIVERSE**
A term for everything that exists, including all matter, space, and time.

**U. T. (UNIVERSAL TIME)**
A measure of time linked to the Sun's apparent daily motion, and which serves as the basis for civil timekeeping.

**VARIABLE STAR**
A star whose physical properties, such as brightness, vary with time. This is usually due to pulsations in the star, but some variables are close binaries in which one star periodically eclipses the other. See also Cepheid Variable, Mira Variable.

**WHITE DWARF**
A small, dense star with a mass similar to that of the Sun, but only about 1 percent of the Sun's diameter. White dwarfs are the shrunken remains of stars that have used up all their fuel.

**ZODIAC**
The band of the celestial sphere either side of the ecliptic through which the Sun, Moon and planets are seen to move throughout the year. Traditionally there were 12 constellations of the zodiac dating from ancient times; the present, officially agreed constellation boundaries mean that the Sun and planets pass through a 13th constellation, Ophiuchus.

# INDEX

# ACKNOWLEDGMENTS

**The author and publisher** would like to thank the following people for their invaluable help in the preparation of this book: Robin Scagell for advice on photography; the staff of Her Majesty's Nautical Almanac Office, Cambridge (particularly Catherine Hohenkerk and Andrew Sinclair) for their providing the information for star charts and planet locators; Jonathan Shanklin for information on comets, Ian Ridpath for technical advice, and David Hughes.

**Dorling Kindersley** would also like to thank Broadhurst Clarkson & Fuller Ltd Telescope House, 63 Farringdon Road, London EC1M 3JB for supplying binoculars and telescopes and in particular Peter Gallon for his invaluable help and advice on these instruments; Francesca Agati, Noel Dockstader David Douglas, Chacasta Pritlove, Ellen Hughes, Owen Hughes, Lara Maiklem, Simon Oon, Ben Raven and Richard Shellabear for acting as models.

**The Author** would like to thank the many people who have helped to produce *New Astronomer*. Special thanks go to Heather Jones and Vanessa Hamilton for their tireless interest, care, and hard work, and to my husband, David, and our children, Ellen and Owen, for their love and support.

**Illustration credits**
All constellation charts plotted by Her Majesty's Nautical Almanac Office, Royal Greenwich Observatory.
Illustrations were prepared by the following:
Julian Baum
Gavin Dunn
John Egan
Jason Little
Shadric Toop
Martin Woodward

**Photography credits**
The publisher would like to thank the following for their kind permission to reproduce their photographs:

a=above;c=center;b=below;l=left;r=right;t=top

Anglo-Australian Observatory: 89 cra, 89 ca, 90 -91, Royal Observatory, Edinburgh 91 tr; Bruce Coleman Ltd :Keith Nels Swenson 73 tl; Robert Dalby: 97 cra; Galaxy Picture Library: Adrian Catterall 62 cb; Alan Heath 46 cb; Bob Garner 32 ca, 50 clb, 54 bl, 58 bl; Bob Mizon 85 cra; Chris Livingstone 90 bc, 103 bc; David Cortner 16 bc; David Graham 42 br, 42 crb, 46 cra, 47 cla, 47 cl, 47 c, 47 cr, 55 bc, 55 clb; David Ratledge 111 tc, 122 cl; Duncan Radbourne 67 br, 86 bc; Eric Hutton 36 c, 66 -67;Gordon Garradd 8 cla, 11 ca, 100 bc, 101 crb, 101 bc, 103 cra, 103 cl; Howard Brown-Greaves 71 crb, 79 cr, 91 br, 106 bl, 109 cl;J.Gillett 112 cl; Jim Henderson 118 bl; John Gillett 110 cl, 113 br; JPL 51 crb, 51 c; Martin Mobberley 67 c; Michael Maunder 11 cl, 71 bl, 75 cr; Michael Stecker 5 cr, 5 tl, 36 bc`, 37 bl, 69 c, 74 -75, 76 -77, 83 cra, 88 cla, 88 c, 88 -89, 88 bl, 89 bl, 93 cl, 93 tl, 97 tc, 104 clb, 105 br, 105 cra, 105 bl, 105 tl, 107 ca, 108 cra, 112 cr, 119 tc, 125 cra, 129 br, 131 tc, 131 cb; Michael Strecker 87 tc; NASA 42 clb, 78 tr;Nik Szymanek and Ian King 92 bc, 111 cb, 113 cra; Paul Stephens 65 tc; Paul Sutherland 77 ca; Robin Scagell 1c, 5 crb, 5 cra, 9 br, 9 tr, 10 C, 12 bc, 12 br, 12 bl, 13 crb, 13 cr, 13 br, 16 cr, 17 tl, 24 cb, 24 cra, 24 crb, 25 tr, 25 cla, 27 cb, 27 cbl, 27 cbii, 27 clb, 27 cbr, 36 ca, 38 br, 39 tr, 39 bl, 42 c, 42 cla, 43 bl, 46 CL, 46 clb, 46 c, 47 tr, 50 c, 50 cla, 54 tr, 54 cla, 54 c, 58 c, 58 cl, 59 tr, 62 c, 62 bl, 62 cla, 63 cl, 63 bc, 63 crb, 63 cra, 64 cla, 66 cr, 66 cl, 66 c, 71 tl, 75 bl, 79 tc, 79 tl, 79 tr, 79 bc, 82 clb, 82 cl, 82 -83 tc, 82 bl, 83 cr, 83 cr, 84 c,84 bc, 84 tr, 84 cr, 85 cl, 85 c, 85 tc, 86 crb, 86 ca, 86 tr, 87 bl, 87 bc, 88 br, 92 tc, 96 bc, 97 crb, 98 cr, 98 c, 99 tl, 99 bl, 99 bc, 100 cl, 101 tr, 102 clb, 103 bl, 104 bc, 106 c, 107 clb, 109 cr, 110 bc, 112 bc, 113 cl, 115 cr, 116 cr, 117 br, 117 cl, 123 bc, 123 cr, 124 crb, 125 crb, 125 cla, 126 bl, 127 tr, 128 bl, 128 CA, 129 cl, 129 cr, 129 tr, 132 bl, 132 cb, 133 bl, 133 cla, 135 tc; Stephen Fielding 122 bc; Steve Smith 123 tc; STScI 11 crb; Tatsuo Nakagawa 13 tr; Y.Hirose 17 cb, 70 cl, 93 cra, 111 cr; Jim Henderson :5 br, 72 -73, 73 cr, 73 cra, 73 crb, 73 br; Jet Propulsion Laboratory: Dr .Robert Leighton 50 cb; Stuart A.Long: 68 crb; N.A.S.A: 5 tr, 10 tc, 10 -11 cra, 11 tr, 32 c, 36 br, 36 cl, 59 cr, 59 cb, 67 tr, 70 -71, 127 cb, 131 cr, JPL 10 cr, 55 tc; Science Photo Library: David Nunuk 6 -7; Frank Zullo 2 -3, 80 -81;John Foster 34 -35; John Sanford 75 tr; Pekka Parviainen 18 -19 tc.

**Jacket design**
by Elaine Monaghan

**Index**
prepared by Hilary Bird

**Design Assistance**
Austin Barlow
Kirsten Cashman
Elaine Hewson

**Editorial Assistance**
Marian Broderick
Samantha Gray
Teresa Pritlove